Growing and Breeding of Potatoes

Growing and Breeding of Potatoes

Sunil Mekala

RANDOM PUBLICATIONS
NEW DELHI (INDIA)

Growing and Breeding of Potatoes

ISBN 978-93-5111-779-7

Published in 2016 in India by

RANDOM PUBLICATIONS

4376-A/4B, Gali Murari Lal, Ansari Road
New Delhi-110 002
Phone : +9111-43580356, 011-23289044, 011-43142548
e-mail: sales@randompublications.com,
info@randompublications.com, randomexports@gmail.com

Reprinted 2026

Type Setting by : Friends Media, Delhi-110089

Preface

The potato is a starchy, tuberous crop from the perennial nightshade Solanum tuberosum L. The word "potato" may refer either to the plant itself or to the edible tuber. In the Andes, where the species is indigenous, there are some other closely related cultivated potato species. Potatoes were introduced outside the Andes region approximately four centuries ago, and have since become an integral part of much of the world's food supply. It is the world's fourth-largest food crop, following maize, wheat, and rice. Modern potato varieties usually have in their ancestry's primitive cultivars or wild species introduced during the present century and may superficially appear to be of very diverse origins. However the breeding procedures used have necessarily involved repeated back-crossing to existing stocks to eliminate the 'primitive' characters of the introductions, and generally 80 % or more of the genes of a variety trace to varieties grown early in this century. In this book, an attempt is made to assess the composition of the gene-pool represented by the early 20th century varieties in relation to that available in Latin America, whence our potatoes came, by surveying the history of introductions and breeding procedures.

– Author

Contents

1

The Systems of Breeding

PLANT BREEDING

One plant may be more favourably situated, getting more light, more manure or water, and because of this is bigger, leafier, or more productive. Another, less well favoured, suffering from paucity of supplies, imbalance of nutrients, or disease, is smaller or deformed or discoloured. The science of genetics, or heredity, is concerned with questions involved in the transmission, from one generation o the next, of all the attributes which gives each its individuality. In short, genetics seeks to explain why sometimes "like begets like," but at other times "begets the unlike." There are two classes of factors which condition the appearance and behaviour of the individual. On the one hand, there are the factors of the environment.

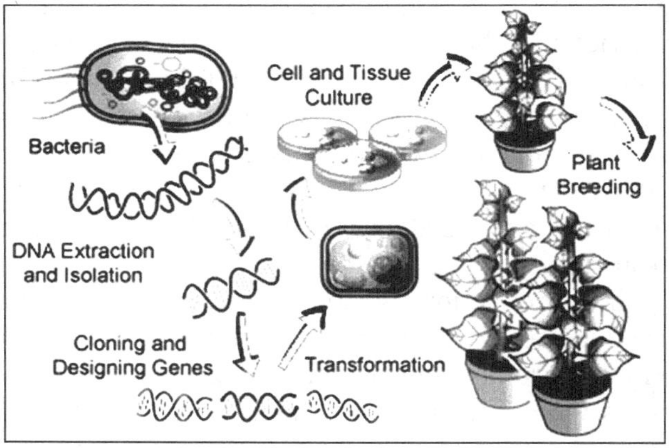

Fig. Plant Breeding

The two plants, had they been grown under similar circumstances, might have appeared and behaved alike. On the other hand, there are differences due to inheritance. There are tall varieties and dwarf varieties of peas; wheat varieties with either white grain or red grain; clover with hairy leaves or non-hairy leaves. The production of these characters is not due to factors of the environment, but to factors inside the plant which were transmitted to it from its parents. The characters of the plant, then, as we see them are the result of an interaction between the factors of environment and the internal, hereditable, or genetical factors. The action of the external factors is' temporary, and produces only *fluctuations* in the course of development; the internal factors are passed on to the next generation unaltered.

The genetical factors are quasi-permanent units passing from generation to generation, and conditioning in each, definite *variations* between individuals differently endowed. Where genetical questions are involved comparisons between plants should be made only between individuals which have been treated alike, and therefore where variation and not fluctuation has been the cause of any differences between them.

Another point which must be clarified is the difference between propagation and reproduction. Propagation is the re-formation of a whole plant from a portion of the body of the "parent." Only body (somatic) cells are involved; only mitotic or somatic cell divisions occur; there is no fusion of nuclei. Hence, all the cells of the "daughter" plants are constituted or endowed exactly as the cells of the "mother," and contain the same set of factors. No matter whether the propagation is natural, as by runners, stolons, bulbs, corms, etc., or artificially produced by cuttings, grafts, buds, it involves only the splitting up of one body into parts and the establishment of these parts as separate entities.

PROPAGATION

By propagation, *one individual* is distributed in time and space. An apple variety, "James Grieve," is bred only once. One tree is grown from a true seed. Every unit of "James Grieve"in the world is merely a part of the original tree, or a part of a part of it grafted on to appropriate roots. Similarly, with a commercial variety of potato, the variety is bred once, then tubers (branches of the body) are distributed. Units produced by propagation are called propagants, or clones. They are identical genetically, and, subject to fluctuation, identical in appearance.

Conversely, in reproduction the body cells of the parent do not enter into the process directly; meiotic division comes between the somatic cell and the gamete, which in its chromosomes carries the genetical factors.

The gamete is not constituted like the parental cell, though it derives its constitution from it. Further, in reproduction there is fusion of two gametes at fertilization to give the zygote from which the body of the new generation derives. The two gametes which fuse may have come each from parents of quite different genetical constitution, and so the offspring of the mating may not wholly resemble either parent but each in part.

Excluding the question of mutation which will be discussed later, propagation gives complete uniformity in the progeny; reproduction holds out the chance of variation.

GENETICS OF PLANT BREEDING

Plant breeding is the art and science of changing the genetics of plants for the benefit of mankind. Plant breeding can be accomplished through many different techniques ranging from simply selecting plants with desirable characteristics for propagation, to more complex molecular techniques

Plant breeding has been practiced for thousands of years, since near the beginning of human civilization. It is now practiced worldwide by individuals such as gardeners and farmers, or by professional plant breeders employed by organizations such as government institutions, universities, crop-specific industry associations or research centers. International development agencies believe that breeding new crops is important for ensuring food security by developing new varieties that are higher-yielding, resistant to pests and diseases, drought-resistant or regionally adapted to different environments and growing conditions.

DOMESTICATION

Plant breeding in certain situations may lead to the domestication of wild plants. Domestication of plants is an artificial selection process conducted by humans to produce plants that have more desirable traits than wild plants, and which renders them dependent on artificial (usually enhanced) environments for their continued existence. The practice is estimated to date back 9,000-11,000 years. Many crops in present day cultivation are the result of domestication in ancient times, about 5,000 years ago in the Old World and 3,000 years ago in the New World. In the Neolithic period, domestication took a minimum of 1,000 years and a maximum of 7,000 years. Today, all of our principal food crops come from domesticated varieties. Almost all the domesticated plants used today for food and agriculture were domesticated in the centers of origin. In these centers there is still a great diversity of closely related wild plants, so-called crop wild relatives, that can also be used for improving modern cultivars by plant breeding.

A plant whose origin or selection is due primarily to intentional human activity is called a cultigen, and a cultivated crop species that has evolved from wild populations due to selective pressures from traditional farmers is called a landrace. Landraces, which can be the result of natural forces or domestication, are plants (or animals) that are ideally suited to a particular region or environment. An example are the landraces of rice, *Oryza sativa* subspecies *indica*, which was developed in South Asia, and *Oryza sativa* subspecies *japonica*, which was developed in China.

CLASSICAL PLANT BREEDING

Classical plant breeding uses deliberate interbreeding (crossing) of closely or distantly related individuals to produce new crop varieties or lines with desirable properties. Plants are crossbred to introduce traits/genes from one variety or line into a new genetic background. For example, a mildew-resistant pea may be crossed with a high-yielding but susceptible pea, the goal of the cross being to introduce mildew resistance without losing the high-yield characteristics. Progeny from the cross would then be crossed with the high-yielding parent to ensure that the progeny were most like the high-yielding parent, (backcrossing). The progeny from that cross would then be tested for yield and mildew resistance and high-yielding resistant plants would be further developed. Plants may also be crossed with themselves to produce inbred varieties for breeding.

Classical breeding relies largely on homologous recombination between chromosomes to generate genetic diversity. The classical plant breeder may also makes use of a number of *in vitro* techniques such as protoplast fusion, embryo rescue or mutagenesis to generate diversity and produce hybrid plants that would not exist in nature.

Traits that breeders have tried to incorporate into crop plants in the last 100 years include:

1. Increased quality and yield of the crop
2. Increased tolerance of environmental pressures (salinity, extreme temperature, drought)
3. Resistance to viruses, fungi and bacteria
4. Increased tolerance to insect pests
5. Increased tolerance of herbicides

BEFORE WORLD WAR II

Intraspecific hybridization within a plant species was demonstrated by Charles Darwin and Gregor Mendel, and was further developed by geneticists and plant breeders. In the United Kingdom in the 1880s, it was the pioneering work of Gartons Agricultural Plant Breeders. In the early 20th century, plant breeders realized that Mendel's findings on the non-random nature of inheritance could be applied to seedling populations produced through deliberate pollinations to predict the frequencies of different types.

From 1904 to World War II in Italy Nazareno Strampelli created a number of wheat hybrids. His work allowed Italy to increase hugely crop production during the so called "Battle for Grain" and some varieties was exported in foreign countries, as Argentina, Mexico, China and others. After the war, the work of Strampelli was quickly forgotten, but thanks to the hybrids he created, Norman Borlaug was able to move the very first steps of the Green Revolution. In 1908, George Harrison Shull described heterosis, also known as hybrid vigour.

Heterosis describes the tendency of the progeny of a specific cross to outperform both parents. The detection of the usefulness of heterosis for plant breeding has led to the development of inbred lines that reveal a heterotic yield advantage when they are crossed. Maize was the first species where heterosis was widely used to produce hybrids.

By the 1920s, statistical methods were developed to analyze gene action and distinguish heritable variation from variation caused by environment. In 1933, another important breeding technique, cytoplasmic male sterility (CMS), developed in maize, was described by Marcus Morton Rhoades. CMS is a maternally inherited trait that makes the plant produce sterile pollen. This enables the production of hybrids without the need for labor intensive detasseling. These early breeding techniques resulted in large yield increase in the United States in the early 20th century. Similar yield increases were not produced elsewhere until after World War II, the Green Revolution increased crop production in the developing world in the 1960s.

AFTER WORLD WAR II

Following World War II a number of techniques were developed that allowed plant breeders to hybridize distantly related species, and artificially induce genetic diversity.When distantly related species are crossed, plant breeders make use of a number of plant tissue culture techniques to produce progeny from otherwise fruitless mating. Interspecific and intergeneric hybrids are produced from a cross of related species or genera that do not normally sexually reproduce with each other.

These crosses are referred to as *Wide crosses*. For example, the cereal triticale is a wheat and rye hybrid. The cells in the plants derived from the first generation created from the cross contained an uneven number of chromosomes and as result was sterile. The cell division inhibitor colchicine was used to double the number of chromosomes in the cell and thus allow the production of a fertile line. Failure to produce a hybrid may be due to pre- or post-fertilization incompatibility. If fertilization is possible between two species or genera, the hybrid embryo may abort before maturation. If this does occur the embryo resulting from an interspecific or intergeneric cross can sometimes be rescued and cultured to produce a whole plant. Such a method is referred to as *Embryo Rescue*. This technique has been used to produce new rice for Africa, an interspecific cross of Asian rice *(Oryza sativa)* and African rice *(Oryza glaberrima)*.

Hybrids may also be produced by a technique called protoplast fusion. In this case protoplasts are fused, usually in an electric field. Viable recombinants can be regenerated in culture. Chemical mutagens like EMS and DMS, radiation and transposons are used to generate mutants with desirable traits to be bred with other cultivars - a process known as *Mutation Breeding*. Classical plant breeders also generate genetic diversity within a species by exploiting a process called somaclonal variation, which occurs in plants produced from tissue culture, particularly plants derived from callus. Induced polyploidy, and the addition or removal of chromosomes using a technique called chromosome engineering may also be used.When a desirable trait has been bred into a species, a number of crosses to the favoured parent are made to make the new plant as similar to the favoured parent as possible.

Returning to the example of the mildew resistant pea being crossed with a high-yielding but susceptible pea, to make the mildew resistant progeny of the cross most like the high-yielding parent, the progeny will be crossed back to that parent for several generations. This process removes most of the genetic contribution of the mildew resistant parent. Classical breeding is therefore a cyclical process.

With classical breeding techniques, the breeder does not know exactly what genes have been introduced to the new cultivars. Some scientists therefore argue that plants produced by classical breeding methods should undergo the same safety testing regime as genetically modified plants. There have been instances where plants bred using classical techniques have been unsuitable for human consumption, for example the poison solanine was unintentionally increased to unacceptable levels in certain varieties of potato through plant breeding. New potato varieties are often screened for solanine levels before reaching the marketplace.

MODERN PLANT BREEDING

Modern plant breeding uses techniques of molecular biology to select, or in the case of genetic modification, to insert, desirable traits into plants.

STEPS OF PLANT BREEDING

The following are the major activities of plant breeding;

1. Creation of variation
2. Selection
3. Evaluation
4. Release
5. Multiplication
6. Distribution of the new variety

MARKER ASSISTED SELECTION

Sometimes many different genes can influence a desirable trait in plant breeding. The use of tools such as molecular markers or DNA fingerprinting can map thousands of genes. This allows plant breeders to screen large populations of plants for those that possess the trait of interest. The screening is based on the presence or absence of a certain gene as determined by laboratory procedures, rather than on the visual identification of the expressed trait in the plant.

REVERSE BREEDING AND DOUBLED HAPLOIDS (DH)

A method for efficiently producing homozygous plants from a heterozygous starting plant, which has all desirable traits. This starting plant is induced to produce doubled haploid from haploid cells, and later on creating homozygous/ doubled haploid plants from those cells. While in natural offspring genetic recombination occurs and traits can be unlinked from each other, in doubled haploid cells and in the resulting DH plants recombination is no longer an issue. There, a recombination between two corresponding chromosomes does not lead to un-linkage of alleles or traits, since it just leads to recombination with its identical copy.

Thus, traits on one chromosome stay linked. Selecting those offspring having the desired set of chromosomes and crossing them will result in a final F1 hybrid plant, having exactly the same set of chromosomes, genes and traits as the starting hybrid plant. The homozygous parental lines can reconstitute the original heterozygous plant by crossing, if desired even in a large quantity. An individual heterozygous plant can be converted into a heterozygous variety (F1 hybrid) without the necessity of vegetative propagation but as the result of the cross of two homozygous/doubled haploid lines derived from the originally selected plant. patent

GENETIC MODIFICATION

Genetic modification of plants is achieved by adding a specific gene or genes to a plant, or by knocking down a gene with RNAi, to produce a desirable phenotype. The plants resulting from adding a gene are often referred to as transgenic plants. If for genetic modification genes of the species or of a crossable plant are used under control of their native promoter, then they are called cisgenic plants.

Genetic modification can produce a plant with the desired trait or traits faster than classical breeding because the majority of the plant's genome is not altered. To genetically modify a plant, a genetic construct must be designed so that the gene to be added or removed will be expressed by the plant. To do this, a promoter to drive transcription and a termination sequence to stop transcription of the new gene, and the gene or genes of interest must be introduced to the plant.

A marker for the selection of transformed plants is also included. In the laboratory, antibiotic resistance is a commonly used marker: Plants that have been successfully transformed will grow on media containing antibiotics; plants that have not been transformed will die. In some instances markers for selection are removed by backcrossing with the parent plant prior to commercial release.

The construct can be inserted in the plant genome by genetic recombination using the bacteria *Agrobacterium tumefaciens* or *A. rhizogenes*, or by direct methods like the gene gun or microinjection. Using plant viruses to insert genetic constructs into plants is also a possibility, but the technique is limited by the host range of the virus. For example, Cauliflower mosaic virus (CaMV) only infects cauliflower and related species. Another limitation of viral vectors is that the virus is not usually passed on the progeny, so every plant has to be inoculated.

The majority of commercially released transgenic plants are currently limited to plants that have introduced resistance to insect pests and herbicides. Insect resistance is achieved through incorporation of a gene from *Bacillus thuringiensis* (Bt) that encodes a protein that is toxic to some insects. For example, the cotton bollworm, a common cotton pest, feeds on Bt cotton it will ingest the toxin and die. Herbicides usually work by binding to certain plant enzymes and inhibiting their action.

The enzymes that the herbicide inhibits are known as the herbicides *target site*. Herbicide resistance can be engineered into crops by expressing a version of *target site* protein that is not inhibited by the herbicide. This is the method used to produce glyphosate resistant crop plants. Genetic modification of plants that can produce pharmaceuticals (and industrial chemicals), sometimes called *pharmacrops*, is a rather radical new area of plant breeding.

PARTICIPATORY PLANT BREEDING

The development of agricultural science, with phenomenon like the Green Revolution arising, have left millions of farmers in developing countries, most of whom operate small farms under unstable and difficult growing conditions, in a precarious situation.

The adoption of new plant varieties by this group has been hampered by the constraints of poverty and the international policies promoting an industrialized model of agriculture. Their response has been the creation of a novel and promising set of research methods collectively known as participatory plant breeding.

Participatory means that farmers are more involved in the breeding process and breeding goals are defined by farmers instead of international seed companies with their large-scale breeding programmes.

Farmers' groups and NGOs, for example, may wish to affirm local people's rights over genetic resources, produce seeds themselves, build farmers' technical expertise, or develop new products for niche markets, like organically grown food.

CROSSING

The genetical factors which pass from generation to generation must be carried in the gametes, the particular nuclei seen in the pollen grain and embryo sac. To effect a cross between two plants, suipollen must be transferred from the stamens of one to the stigma of the other, precautions being taken that no other pollen is present. All being well, the pollen germinates and fertilization follows.

The two plants so "crossed"constitute the "first parental generation," or Pi, and their immediate progeny, the "first filial generation," or Fi. If the pollen supply used in fertilizing the Fi is restricted to its own spores, and it becomes self-fertilized or "selfed," the next generation is known as Fs.

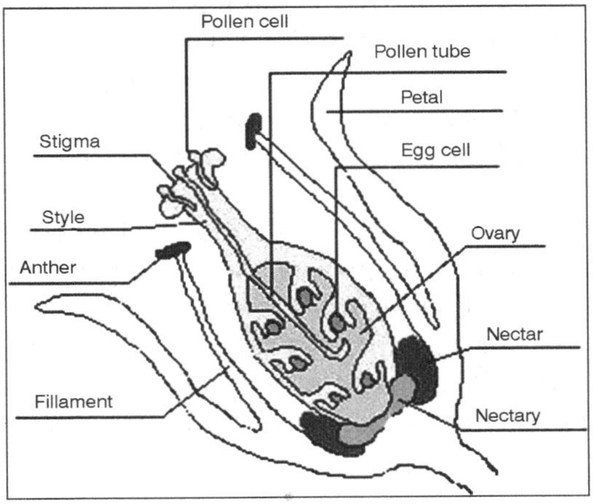

Fig. Pollen Germinates

Simple Cross

By appropriate crossing and selfing the behaviour in heredity of the different factors may be followed. A very simple case is seen in wheat. Some varieties have the spikelets arranged on the rachis, with an interval between. In others, the spikelets are placed quite close together and the "head"is compact. The one is said to have a "lax head," the other a "dense head." If two plants, one from each of these types, is crossed, the Fi is intermediate as regards this character. When the Fi is selfed, the F2 shows three classes of plant.

In one class all the plants have lax heads; in the second they are all dense; while in the third they resemble the Fi, and are intermediate in character. The cross is made either way, the dense head type or the lax head being used as the female, it matters not; if sufficient plants are grown in Fs, the numbers in the three classes will always bear the same numerical relationship to each other: 25 per cent, lax, 50 per cent, intermediate, 25 per cent, dense; or the ratio 1:2:1.

DOMINANCE

When other characters are considered in a hybridizing experiment, the Fi may not be intermediate but resemble one of the parents. There are bearded wheats and non-bearded. That is, the "chaff" which envelopes the grain may have a long tapering awn (bearded), or this may be absent (beardless). When a bearded wheat is crossed with a beardless one, the Fi is bearded. The gene for bearded dominates in the Fi. It is said to be dominant to the non-bearded gene, which conversely is said to be recessive. In Fs only two classes appear—beardless and bearded. The numbers of plants in ttyese classes are in the ratio of 3: 1. In Plate 120 specimens from the actual plants in such a cross have been mounted in such a way as to illustrate the behaviour seen in a mating of this type. In the lowest line of plants (the Fa) it will be seen that there are twelve bearded plants to four without beards; that is, the numbers in the two classes are in the ratio of 3: 1. This is quite simply interpreted when it is remembered that one "dose"of bearded "swamps"or masks the presence of one dose of beardless.

It is clear that the 3: 1 ratio is merely the 1:2:1 with the first two classes amalgamated by the effect of dominance. Only one plant (the homozygote) in every three of the bearded class will breed true. In this particular case the heterozygote produces much smaller awns than those borne by the homozygote. This partial production of the character is called partial dominance. When the heterozygote is quite similar to the homozygote the dominance is said to be total.

Chromosome Gene

A plant shows hundreds of units of character and must therefore have hundreds of pairs of genes, yet it has only a limited number of chromosomes in each nucleus; in bread wheat there are twenty-one pairs in the soma or twenty-one individuals in the gamete. Many other plants have fewer chromosomes, yet these bodies have to carry all the genes. It is known that each chromosome carries a number of genes, and these are distributed up and down the chromosome in linear order like beads on a string.

Each chromosome always carries the same gene set. The distance between neighbour genes on a chromosome is constant, and the order in which they occur from one end to the other never varies in normal cases. Linkage would prohibit the possibility of new combinations being bred, were it not for the fact that at syndesis in meiosis, the chromosomes of homologous pairs become closely intertwined, and in separating give the appearance of having broken and united with an exchange of parts between the two members of a chromosome pair. The points where they cross-over each other in the intertwining, called a *chiasma,* can be seen under the microscope.

On the left is the original chromosome pair; next the two intertwined; next the two chromosomes are shown reformed with exchange of parts, while on the extreme right are the chromosomes as they go forward to complete mieosis and pass to the gametes.

Phenotype and Genotype

A plant described on the basis of its external appearance is called the phenotype. If described by reference to the genes inside it is called the genotype. In the Fs, then, there are bearded phenotypes, some of which (the heterozygotes) are genotypically bearded and beardless, while the others (homozygotes) are phenotypically and genotypically beardless.

Genetical Purity

In order to distinguish clearly between a homozygous and a heterozygous bearded, or, in other words, to distinguish between the two possible genotypes, the plants must be self-fertilized and bred-on. The progeny of the two genotypes will differ. On self-fertilization the homozygote produces only one class of plant, or, in other words, it "breeds true." On the other hand, the heterozygote segregates. This "proving"of the plant by reference to the character of its offspring is called the progeny test. Where only a small number of plants can be grown, it may so chance that following on self-fertilization the recessive form does not happen to be produced.

If, however, the suspected heterozygote, instead of being self-fertilized, is crossed with the pure recessive parental type, the chances of the recessive condition appearing in the progeny is increased. Every pollen grain from a beardless plant carries the gene beardless. This, supplied to a heterozygous plant, ensures that every ovule carrying the recessive gene will produce a beardless individual. In short, the cross of a heterozygote to the recessive ensures a 50/50 distribution in the two classes. Crossing back to the recessive parental type is a special case of progeny testing called the *back-cross.*

The phenomenon of segregation and recombination of these genetical factors was discovered in the garden pea by Mendel. The ratios which he demonstrated are called Mendelian ratios, and the study of their various forms, Mendelism. Very many other but similar cases have been described in all sorts of organisms by many investigators. Sometimes dominance appears, sometimes the Fi is intermediate. When the parents of a cross differ in two pairs of characters, both showing dominance, four classes of plant appear.

This is due to two 3:1ratios being superimposed. Re-examination of the photograph will show that the left-hand awnless parent (variety Iron) has also white chaff while the right-hand bearded parent has red chaff. These two parents differ in two characters—awnedness and chaff colour. The Fi is seen to resemble the "April" parent in that it is both bearded and red-coloured in the chaff. When the Fi proceeds to form gametes the two pairs of genes involved segregate quite independently of each other, and therefore produce equal. Quantities of pollen grains in four different classes, each class differing from the others in gene content. Similarly there are four classes of female gamete.

THE GENE

The mechanism which lies behind this segregation and recombination of factors is simple. It is now established that the genetical "factor"has a material basis in a piece of the structure of a chromosome. The name for this material basis of the theoretical factor is the gene. Each gene is known to occupy a definite place (the locus) in a definite chromosome. The chromosomes are paired bodies, and in a diploid (somatic) cell-nucleus there are two homologues, each carrying the members of the same gene pairs. The chromosomes of each homologous pair at meiosis part company.

This means that every gamete will receive one chromosome from each chromosome pair, and therefore one gene from each gene pair. At fertilization the chromosome pairs, and therefore the gene pairs, are reconstituted. The gene controlling the arrangement of spikelets on the rachis exists in two forms or potentialities. There is one potential for the lax condition, and the other potential for the dense condition. All the gametes of the lax parent will carry the gene for lax (L), and all the gametes of the dense parent will carry the gene for dense (D). These two different aspects of the gene pair meet at the fusion of the crossed gametes, and are therefore present in the somatic nuclei of the Fi, where they exert their influence.

When the Fi in its turn proceeds to form gametes the chromosome homologues part, and therefore the genes they bear part also. One half of the pollen will receive the gene aspect for lax, the other half for dense. Similarly, 50 per cent, of the ovules will carry "lax," and 50 per cent. "dense." On self-fertilization of the Fi, it depends entirely on chance whether a pollen grain carrying "lax"meets an ovule carrying "dense"or "lax." Similarly for pollen grains carrying "dense," as there are equal numbers of dense and lax carriers in both the male and female gametes.

The Heterozygote

These hybrid plants whose somatic nuclei develop from a zygote containing both aspects of the gene pair *(e.g.* lax-producing and dense-producing) are known as heterozygotes, and are said to show the heterozygous condition. On the other hand, plants whose nuclei have two "doses"of the one aspect of the pair are homozygous, and such plants are homozygotes. The homozy-gote breeds true when self-fertilized, for all its gametes are alike. The heterozygote when self-fertilized shows segregation.

When it forms pollen and ovules the two aspects of the gene pair part company and, depending how they meet for fertilization, produce plants of different character. The essential point here is that gene pairs segregate at reduction division because their chromosomes segregate. They recom-bine at fertilization because at that point in the life-cycle the chromosome pairs are reconstituted. This explains why there are three classes in F2, and the fact that when large numbers are grown in this second filial generation the numbers in each class occur with the bounds of reassortment on the basis of pure chance in the ratio of 1: 2: 1.

PLANT MUTATION

It may be asked now how a plant comes to have two aspects of a character or how a character pair comes to be. It is known that the wild-type gene as it occurred originally may alter its nature. The alteration produces a new expression of the character.

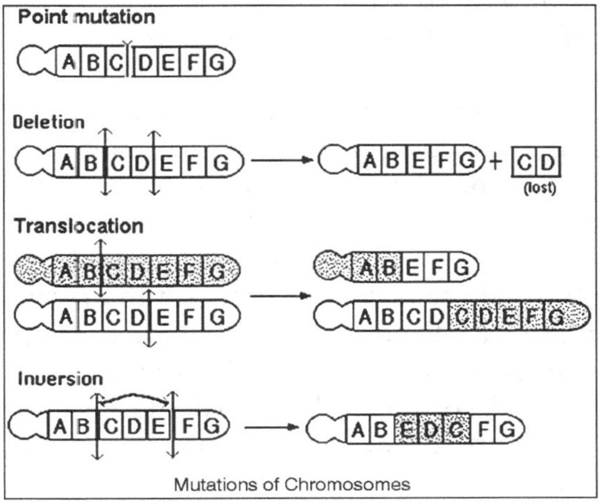

Fig. Mutation

The process, or act of alteration, is called mutation. The new type of plant so produced is called a mutant or a mutation. A gene in the majority of cases mutates only once, and thus gives a character pair, the wild type and the mutant. The same gene, however, may mutate again and give three forms of the gene—the original wild type and two mutants. This provides three expressions of the character. Only two of these three can be present in any diploid somatic nucleus, and only one in any gamete. A case may be taken from the haricot bean, plants of which may have green foliage and green pods (Gr) or green foliage and yellow pods (G), or yellow foliage and yellow pods (g). When Gr is crossed with G, the Fi shows green pods and foliage. The Gr condition is dominant to the G condition. When the G type is crossed with the g, the green foliage and yellow pods condition is dominant.

Thus we see a series of decreasing dominance—Gr dominant to G, and either of these dominant to g. Such a graded set is called a multiple-allelomorphic series. All sorts of quantitative and physiological conditions are known where the genes for the characters belong to such a series. In some series there may be more than three aspects, for cases are known where the gene has mutated five or six times, and six or seven character expressions exist.

ASPECT OF CYTOLOGY IN HETEROPLOID

Another aspect of cytology affecting Mendelian ratios must now be referred to. It sometimes happens that at mitotic or meiotic division the daughter nuclei do not separate cleanly in the usual manner. In these cases the chromosomes, either as a whole set or as part of a set or even individual members, lag behind. The laggards which should have gone to one daughter nucleus are included in the other. When this occurs, nuclei are formed with a number of chromosomes different from that normally found in the species. A nucleus with such an aberrant number of chromosomes or a plant in whose nuclei they occur is said to be heteroploid. If a whole chromosome set is replicated, the heteroploid is said to be balanced or *euploid.* If only a part of the set is replicated, the nucleus is of the unbalanced or *aneuploid* class. Most attention is directed to those cases where whole sets are concerned, that is, to the euploid form.

Where one replication of the normal diploid has taken place, and the homologues are in threes, the plant is said to be *triploid.* Where the diploid has doubled and the homologues are in fours, the plant is said to be *tetraploid.* Plants are known where the body cells contain only half the normal diploid number; that is, the number normal to the gamete of the species. These plants are known as *haploids.* There is thus an ascending series in the terminology: haploid, diploid, triploid, tetraploid, pentaploid, hexaploid, and upwards. These have been formed in nature as the result of some unknown stimulus, or artificially in cultivation by the use of such drugs as colchicine.

HETEROPLOID'S EFFECT

Giantism

Replication of chromosome sets often results in the production of a bigger (giant) form of plant. This in itself may be very useful as, when it occurs in a plant of a forage crop.

Sterility and Fertility

Heteroploidy may be very useful in regard to sterile hybrids. These hybrids often result from crosses between plants which are not very closely related, but close enough to form a new hybrid plant. The chromosomes derived from the two parents are sufficiently homologous to form a mixed nucleus in Fi, but not so as to behave normally at meiosis. The hybrid fails to form viable gametes. Such hybrids or "mules"are sterile, and therefore, unless they can be propagated successfully and economically, are of little or no use in practice. Replication of the mixed sets of chromosomes, if it occurs in such a sterile hybrid, may render it fertile.One of the clearest examples of this is furnished by a non-agricultural plant.

Primula verticillatay with nine pairs of chromosomes in the somatic nuclei, can be crossed with Primula floribunda, which also has nine pairs of chromosomes.The gametes of each, therefore, contain nine chromosomes, and these form the hybrid. None of the nine floribunda chromosomes are homologous with the nine verticillata members, and therefore successful meiosis in the hybrid is barred. Viable gametes are not possible, and the plant is sterile. The sterile hybrid formed a euploid and immediately became fertile. Schematically, the mating, the chromosomes being symbolized by the initial letter of the species from which they are derived.By replication, each V chromosome obtains a homologue and similarly each F chromosome. Normal meiosis is possible and gamete production follows. Many species not closely related can form hybrids which are sterile.

It is probable that many of these will be rendered fertile when duplication occurs. Wide crosses as between, a wild grass like couch and wheat, or between turnip and cabbage, are important because the new gene complex brought into being results in more than a mere addition of the characters of one parent to those of the other. Genes even in a normal diploid soma interact on each other to some extent, and when two complete gene complexes interact the effect is greatly increased, so that a new plant different from both parents is brought into being.

This is paralleled in chemistry when, the union of sodium, a white silvery metal, with chlorine, a green gas, produces a new substance, sodium chloride with characters quite different from the parent substances. The formation of heteroploids has occurred in nature. Our modern bread wheat as it was evolved is a complex hexaploid of diverse parentage with three sets of seven, that is, twenty-one, chromosomes in the gamete, and twenty-one pairs (forty-two individuals) in the somatic nucleus. One interesting result of heteroploidy occurs when the component sets of chromosomes each contribute to the new plant similar genes.

A case in point arises in connection with the production of the red colour in the grain of bread wheat. Here, if an appropriate variety with red grain is crossed with a variety having white grain, a 3: 1 ratio red. If a different red-grained variety had been used as a parent, a 15: 1 ratio red: white in Fs would have been obtained. If yet another variety had been employed, the ratio of red: white in F2 would have been 64: 1. Results such as these indicate that three pairs of genes are involved in the production of grain colour.

Furthermore, each pair must be located on different chromosome pairs, because the genes segregate and recombine freely. In short, there must be six chromosomes (three pairs) in each hexaploid somatic nucleus, all of them carrying a grain-colour gene. Cytological evidence shows the bread wheat to be hexaploid. Multiple genes (polymeric series) of this type become important when they affect quantitative characters, and each dose of, the dominant produces an increase in the character. Sugar content of sugar beet depends on a set of replicated genes, and every additional "dose"of gene produces an increase of sugar content.

SELF-FERTILIZATION AND CROSS-FERTILIZATION IN STERILITY

Self-fertilization and cross-fertilization have been referred to. These are often barred by sterility. Almost any cabbage plant will illustrate the point. If its own pollen only is allowed access to its stigmas a cabbage will set few or no seeds. If pollen from other cabbage plants gets access to the stigma, fertility is complete. This proves that its ovules are fully functional. Its pollen is fully effective in producing seed on most others of its fellows. The gametes of the plant accordingly are fully viable but self-incompatible.

It is now known that the tube of a pollen grain carrying any Sallelomorph can hardly pass through a style carrying this same allelomorph in the pair resident in the cells. A similar mechanism affects many varieties of fruit tree. If an orchard has been planted only with such a self-incompatible variety, every one of the trees produced by vegetative propagation will have exactly the same gene complex. Hence, every stigma will contain incompatible genes in common with the pollen supply and fertility will be barred. Such an orchard must be interplanted with a few trees of another variety which is known to be capable of supplying suipollen. Many plants are known to have a cytological constitution which does not permit of normal meiosis. The plants are usually of hybrid origin. Many fruit trees are of this nature and do not form viable gametes.

VIGOUR LOSS

Where sterility is no bar, and continued self-fertilization is practised, another disability may show itself. The vigour of the progeny derived through continued self-pollination falls progressively generation by generation. In maize, after a few generations of self-fertilization the inbred lines become weak, and in extreme cases may fail completely and die out.

Hybrid Vigour

Curiously, when two inbred but unrelated lines are crossed the F1 is often extremely vigourous. Out-breeding brings back vigour. This vigour in a hybrid has been called *heterosis* or *hybrid-vigour.* It shows only in the F1 and does not appear in F2 when the F1 is self-fertilized. When appropriate inbred lines are used as parents, the Fi is often very much more vigourous and yields much more than any commercial lot. Heterozic or Heterozygous maize seed is now sold for crop production. In fact, for this crop, practically all progressive growers sow only hybrid vigourous seed. All of the Fi crop goes for consumption as food, etc.; none is retained for sowing. Each year the Fi is made anew from inbred parental lines which are known to give the maximum effect when crossed.

The commercial production of heterozic maize seed is simple. The two parental lines are interplantcd in rows. At flowering time all the male inflorescences are removed from the plants designed as the seed parent. It is therefore pollinated only by its neighbour, and all the "seed"produced by it is of hybrid origin. In any crop produced by vegetative propagation the vigourous hybrid need be produced only once. All clones arc genetically identical and just parts of the Fi itself. They retain all its characteristics in full effect. This is well seen in the poplar tree used for wood pulp. Two parents arc crossed. If the progeny is highly vigourous, suckering is induced. The suckers are planted out. The trees from such suckers grow very fast and harvest is brought nearer by a number of years.

Fig. Heterosis

SELECTION OF PLANT

Cytological aberration and genetical hybridity provide plant populations variable in character. Variation in a crop, no matter how induced, provides the plant selector with his material. Two inbred strains of Maize on eitiier side of the vigourous hybrid got from their mating. A new combination of genes or gene complexes can be the starting point of a new variety waiting to be selected and multiplied. There are three methods of selection.

Pure Lines

The first consists of selecting single plants, self-fertilizing them, and maintaining the progeny of each as a separate "line." In effect this subjects each selection to the progeny test. The best line which breeds true can be retained and the others scrapped. This method is known as single-plant selection. If the original parent was homozygous for the characters observed, a non-varying or *pure line* will have been formed. A pure line consists of the descendants of one homozygous individual, self-fertilized and never cross-fertilized. This amounts to *genotypic* selection. A line pure for *all* genes is probably not possible, but a line pure in all the genes that matter is obtainable in practice. It is to be noted that, so far as concerns the genes for which the line *is* pure, further selection can have no effect. A pure line, then, by definition is non-varying. Apart from environmental fluctuation and the possibility of mutation, no alteration in appearance or behaviour is possible.The second method of selection is to choose all the plants in a crop which appear to conform to a predetermined standard. Such is mass or *phenotypic* selection. This is often useful with plants which are individually self sterile, or where the character is quantitative and depends on a comparatively large series of polymeric genes.

Mass selection gives an immediate effect with a large bulk in the first generation. As generation follows generation, if re-selection is not practised on each, the effect wears off, and the strain "runs back"to an average. Mass selection is often possible on large-scale lines. In many countries growing crops are inspected officially, and where a crop conforms to a definite standard the grower receives a certificate for it. When such a certification scheme is widely patronized it amounts to a large-scale mass selection.

The third method of selection is not to select out the good plants, but to pick out from a crop the wrong ones or "rogues."These are the plants which do not conform to standard. Roguing has a potent influence when the rogue is due to a dominant gene, and when the desired character is due to a recessive. Roguing out a recessive is not really effective in practice, for it leaves in the crop the heterozygotes. In this latter case it usually pays better to develop a pure line following a progeny test.

REASONS FOR PRIVATISATION OF PLANT BREEDING

Plant breeding can be conceptualised as an investment that develops improved varieties with the potential to generate future benefits in the form of improved crop productivity, reduced costs of production, and/or higher returns. Potential value from improved cultivars will be realised only if and when farmers adopt these cultivars in their cropping systems, AND when consumers willingly purchase the food or other crop products in a competitive market. Growers will only adopt these new varieties if they provide real financial benefits that exceed the costs of adoption, including any additional costs of acquiring the improved variety.

Similarly, consumers will only knowingly purchase food from these new varieties if by so doing they derive a net benefit in the form of enhanced attributes and/or lower prices relative to available alternatives. In common with other forms of investment, the rate of return will depend on the discounted value of the flow of future benefits net of present value of all costs neccessary to generate such benefits. Arguably the most important reason for the growing trend to privatisation of plant breeding has been significant changes in the ability of plant breeders to appropriate at least some of the benefits from improved grain varieties that otherwise would be captured by growers.

Specifically, the incentive for firms to invest in plant breeding depends on the ability to exclude grain growers, and often competing plant breeders as well, from commercial exploitation of a breeder's varieties unless they pay to do so. For some crops such as corn, the development of hybrid technology provided genetic copy protection that enabled plant breeders to capture much of the value from heterosis as well as other superior traits. For other crops, it has been the expanding scope of intellectual property rights that has enabled the capture of some of the value created by plant breeding. So while it has been the application of modern science to plant breeding that has generated much of the potential for value creation in the grain supply chain, it has been extensions to the legal framework for intellectual property rights that have made possible private capture of enough of the value created by plant breeding to provide the private sector with an incentive to invest more in plant breeding.

The most significant of these intellectual property rights are patents and Plant Breeder's Rights. In recent decades, both court decisions and legislative changes have expanded the scope and impact of these two types of intellectual property right appreciably. Complementing these developments in the institutional framework have been scientific discoveries that led to greater potential for value creation by improvements in plant breeding methods. Apart from hybrid technology, other recent advances include new technologies that improve the efficiency of all plant breeding, including both conventional and transgenic plant breeding.

Such techniques include double haploidy, plant regeneration systems, molecular based hybrid technologies, and marker assisted selection. Use of these techniques in conventional plant breeding is already reducing the time lags from initial crosses to release of new varieties. Potentially beneficial outcomes from the application of these technologies to plant breeding include one or more of the following:

- cheaper development of improved crop varieties.
- faster/earlier development of improved crop varieties.
- development of superior improved crop varieties, that are more productive, produce better quality grain, or both.

In addition, there has been the more controversial development of transgenic technologies used to produce GMO's. Potential beneficial outcomes from transgenic technologies include:

- development of improved crop cultivars with novel agronomic/input traits that enable lower average costs of production.
- development of improved crop cultivars with novel quality-enhanced traits for which consumers are willing to pay a price premium.

On the other side of the coin, publicly funded rural research has been under pressure for at least the last two decades. In part, this been due to a growing perception that grain growers have been the primary beneficiaries of the traditional plant breeding programmes. Historically these programmes have been funded mainly from consolidated revenue. To some extent, this concern has been addressed by the relatively recent evolution of the GRDC and similar bodies that rely heavily on collective industry funding to support much of their investments, but the fact remains that a significant part of plant breeding programmes is still publicly funded.

Governments also now demand greater accountability at the same time that they reduce funding for agricultural research and extension. As a result, many "public" institutions are under pressure to become at least partially self-funding, and are starting to charge for selected goods and services. Public research institutions also seek to patent and/or commercialise discoveries made in the course of government funded research, or pursue opportunities to license technologies to the private sector. Public plant breeding programmes have not been immune to government pressure to generate revenue from their activities. Like private business, their capacity to capture a high proportion of the net benefits of new varieties depends on:

- a legal basis to establish ownership of the intellectual property embodied in the variety,
- the capacity to exclude potential users who are not willing to pay the nominated price,
- the costs of monitoring and enforcing compliance,
- the capacity for price discrimination.

Pricing practice by public institutions is still evolving. If they start charging significant fees at levels approaching full cost recovery, and exclude farmers unwilling to pay these fees from access to new varieties, then they cease to be public plant breeding organisations within the meaning of the term in this stage.

Finally, agronomic practice by grain growers has become increasingly sophisticated and much more tactical. In particular, many growers now make decisions about which varieties to grow each season on the basis of the latest possible information about the climatic outlook and other seasonal indicators, such as soil moisture levels as well as weed and disease threats. Consequently they are less likely to use seed saved from the previous harvest, and more likely to purchase new seed of the desired variety from a seed merchant. This change in farming practice will increase the size of the seed market, and improve the economics of private plant breeding.

CLASSICAL PLANT BREEDING FOR PEST RESISTANCE

Wild relatives of crop plants such as beans, wheat, and maize are not uniformly resistant to insect and disease pests. This can be demonstrated in simple fashion—when selections of these wild populations are set out in plant-rows, some of them are highly susceptible, others are resistant, and some are intermediate in resistance to the common pests of the region.

The first plant breeders, those women and men who domesticated crops such as beans, maize, and wheat, could save only those genotypes that had some level of resistance, *i.e.*, those individual plants that did not succumb to pest depredation. In effect, therefore, they selected for pest resistance and thus changed the population structure of their crop species in favour of resistance genes. This change made it possible to grow the crops in monoculture, which was convenient for food production and harvest. It was also convenient for multiplication of disease and insect pests that might not be affected by the limited sample of resistance genes.

Plant breeding thus set the stage for sequential cycles of pest resistance and pest susceptibility of crop plants. We have no direct record of the consequences of this ancient ecological meddling, but myth and historical accounts tell of disastrous disease epidemics and insect outbreaks, so one can assume that from time to time large plantings of crops that were uniformly susceptible to a new kind of insect or disease fostered increases of that pest to epidemic proportions. Resistance genes were essential for crop domestication and monoculture but they did not guarantee perfect safety. We have no record at all and little or no speculation about how the newly domesticated crops might have affected their wild relatives, which no doubt were growing in close proximity to the domesticates.

RESISTANCE BREEDING AFTER MENDEL

Genetics-based plant breeding, launched in the early years of the 20th century, produced new crop varieties with improved resistance to major disease and insect pests. Usually such resistance was developed as a second phase—a rescue operation—after new varieties, selected primarily for high yield, were discovered to be susceptible to a particular insect or disease. Breeders found early on that they could identify single genes (usually dominant) that conferred essentially complete resistance to the pest in question. Varieties containing such excellent resistance were developed and released for large-scale farmer use. But breeders then discovered, all too often, that the "perfect" resistance lost its effectiveness after a few seasons. They soon learned, with the aid of entomologists and plant pathologists, that insect and disease pests are highly diverse genetically, and that almost without fail a rare pest genotype will turn up (or perhaps be created *de novo* by natural mutation) that is not affected by the newly-deployed resistance gene. The new pest genotype multiplies and the crop variety's resistance "breaks down."

As years went by, breeders found that some kinds of resistance did not fail, and that such resistance often was less than complete; the plants suffered some damage but gave satisfactory performance overall. This longer lasting resistance was dubbed "durable" resistance. Further, the breeders discovered that durable resistance usually (but not always) was governed by several genes rather than by one major gene. The multifactorial kind of resistance has been called "horizontal resistance." The major-gene resistance has been called "vertical resistance."

The good news, then, was that breeders could identify and breed for durable resistance. The bad news was that the breeding was more difficult because several genes had to be transferred at one time, thus requiring larger populations for selection, as well as multiplying the usual problems with "linkage drag" (undesirable genes that are tightly linked to the desired ones). To this day, breeders use both kinds of resistance in varying proportions, according to the crop and where it is grown. At first, breeders found and used resistance genes from the adapted, local landrace populations that also were the initial gene pool as a source of resistance genes for their new varieties. As years went by, these gene pools began to dry up and breeders looked further afield, turning to exotic (unadapted) landraces, and even to wild relatives of their crop. Sometimes they made extraordinary efforts to hybridize the domestic crop with a very distant wild relative—making a cross that could not succeed under natural conditions. Embryo rescue and even x-ray treatments were used to make "unnatural" crosses and derive breeding progeny from them. The breeders fooled around with Mother Nature; they moved genes farther than natural processes would allow.

But the breeders as a whole preferred to not breed from exotic varieties or distant and often wild relatives. They used exotic material only when there was no other choice. This preference was due not only to the difficulty of wide hybridization, but also to the fact that exotic germplasm exacerbates the problem of undesirable linkages. Few or none of the foreign genes—except the desired resistance genes—were suitable for the needs of high yielding, locally adapted varieties. But often the breeders had no choice; either they got the needed resistance genes from a distant relative, or they got nothing at all.

At about this time, breeders realized that it would be important to conserve remnant seed of landraces from all around the world, but especially from the centres of diversity of their crop. As farming worldwide grew more commercial, farmers turned more and more to professionally bred varieties that were better suited to commercial production, and in so doing they abandoned their landraces. If remnant seed of those landraces was not collected and saved in special storage facilities, the genetic base for crop breeding in the future would be drastically narrowed. Seed "banks" were needed. Through the efforts of a few far-sighted plant breeders, seed banks were established in several countries and in international research centres.

So at the end of the 20th century, plant breeding for pest resistance had laid out the genetic framework of vertical and horizontal resistance, and identified important sources of new resistance genes, *i.e.*, plant germplasm from anywhere in the world. Sources were limited, however, to the crop species itself or its relatives, either wild or cultivated. All of the introduced genes therefore came from plants. Plant breeders selected not only for tolerance or resistance to disease and insect pests, they also selected for tolerance to abiotic stresses such as heat and drought, cool temperatures, or nutrient imbalance. Much of this selection was involuntary; in selecting varieties with top performance over many seasons and many locations the breeders necessarily selected varieties with tolerance to the prevailing abiotic stresses of the diverse seasons and localities. In selecting for tolerance to environmental stresses, breeders necessarily changed the genetic makeup of the crop species, altering it still further from that of the original wild species, which had been restricted to certain environmental niches. Witness teosinte (the probable parent of maize), restricted to certain habitats in Mexico as compared to maize that now is grown in nearly every country of the world except Iceland. Global distribution of crop plants often means that they are grown with no proximity to wild relatives that might intercross with them. Teosinte is not found in Germany or China, nor for that matter in the US Corn Belt. In other cases, however, wild species with hybridization potential coexist with their cultivated crop relatives, often as weeds. Canola, sunflower, and grain sorghum are examples of crops with hybridization potential with either a related species (canola with wild mustards) or with a weedy form of the same species (sorghum with shattercane, cultivated sunflower with wild sunflower).

VALUE CREATION AND CAPTURE ISSUES

It is inevitable that the growing privatisation and commercialisation of plant breeding in the Australian grains industry will lead to increased competition between plant breeders. While this trend is seen as threatening by some people, there are compelling grounds for viewing this change as an opportunity to create more value for the industry and the nation provided that certain conditions are met. First though the way in which plant breeding creates value is discussed.

In a market economy, the ultimate determinant of the aggregate value of all activities in the food supply chain is the amount that consumers are willing to pay to consume end-products. Value creation occurs when there is an improvement at any step in the supply chain that increases aggregate value in consumption.

Market forces, mediated by the institutional and policy framework, including intellectual property rights, will determine the extent to which value is captured at various stages along the supply chain. In the grains industry, the potential exists for new varieties to create value by lowering the cost of producing and delivering grain and grain products to consumers; and/or by enabling the production of superior grain products for which consumers are willing to pay higher prices. This potential for value creation will be realised when new varieties from the breeding programme are released and adopted by grain growers, and the resulting products are purchased and consumed by end-users. The application of modern science to plant breeding has dramatically increased the potential to create extra value in the grain supply chain. New plant breeding methods such as di-haploidy, embryo rescue, and rapid breeding cycles have sped up the development of new varieties and reduced breeding costs. Furthermore, molecular marker technology enables breeders to be much more selective and effective at identifying desirable traits in germplasm collections and incorporating these traits into elite lines. Transformation technologies have significantly expanded the range of traits that plant breeders can access.

Greater value will be created if plant breeders can produce more and better varieties faster and cheaper. To fully realise this potential for value creation, more rather than less investment in plant breeding is needed, and plant breeders need to take a more selective and strategic approach to plant breeding decisions. For reasons already discussed, less rather than more public funds are likely to be available for investment in plant breeding, so most new investment will need to be privately funded. Funds will be forthcoming only if prospective rates of return from private plant breeding are sufficiently attractive, and this will depend on the extent to which market forces and intellectual property rights allow and enable breeders to capture the value that they create. Innovation involving changes to established practice in plant breeding is essential to take full advantage of the availability of the new breeding technologies.

Sometimes organisations are able to achieve such a transformation in the absence of competitive pressures, but often they cannot. For both public sector organisations and private firms, competition commonly provides the necessary impetus to drive change and risk taking. Notwithstanding the potential benefits from privatisation, some people perceive threats to the interests of grain growers and/or to the national interest. Some threats are more imagined than real. Other threats are real enough for some individuals or groups, but are outweighed by broader community benefits. There also are some legitimate grounds for concern about institutional arrangements that could limit the efficiency gains from a more competitive plant breeding system.

One possible outcome of greater privatisation of plant breeding is that a small number of large multi-national firms crowd out other competition, and that they might then exploit their market power to capture almost all of the value created by plant breeding. Pricing practices by these large life science companies also might threaten future competitiveness of other sectors of the farming industry, as well as limiting widespread value creation. The commercialisation of Bt cotton by Monsanto in the USA and in Australia provides one such example. Bt cotton was commercially introduced in the USA in early 1996, and later the same year in Australia in time for the 1996/97 growing season. Monsanto Corporation is the patent holder for the Bt technology, and sought to appropriate the benefits generated by its intellectual property through a technology fee. In the USA, the initial level of the technology fee charged to cotton growers was US $32/ acre.

In Australia, Monsanto set the technology fee at A$245/ha, which at then current exchange rates translated to about US $70/acre, or approximately double the level set in the USA. Even though Monsanto subsequently introduced some concessions in response to pressure from the domestic industry, Australian cotton growers still ended up paying a considerably higher technology fee than their US counterparts. The considerable monopoly power that the large life science companies possess over proprietary key enabling technologies for plant breeding also is a possible threat to continued technology development in a country such as Australia. Arguably, legal disputes over intellectual property rights that block widespread utilisation and/or further development of the technology pose a major threat to long run value creation from transgenic technologies. Concerns have been expressed about possible patent gridlock, excessive secrecy and duplication of inventive effort, excessive transaction costs to license patented technology, prisoner dilemma type impasses, and/or the "tragedy of the anti-commons".

Such supply side problems might tie up the technology in the courts and block commercial implementation for years, if not decades. It also may threaten the *freedom to operate* for public and local industry research and plant breeding programmes. Given examples such as this, it is not surprising that Australian grain growers have already demonstrated a willingness to fund local plant breeding firms to forestall such a threat, and to ensure ongoing access on reasonable terms to locally bred varieties that maintain Australia's competitive advantage in international grain markets.

Finally, there are real concerns relating to the provision and utilisation of certain key "knowledge rich" inputs to the plant breeding process. Examples of these "knowledge rich" inputs include germplasm collections of land-race and elite breeding lines, and results of pre-breeding research such as agronomy; biometry; entomology; genetic mapping and molecular marker development; germplasm collection and conservation; information and database systems; molecular biology research; plant pathology; plant physiology; plant quarantine; product chemistry; and quantitative genetics. Most such inputs are key enabling technologies for plant breeding that provide the foundation for ongoing long-term variety improvement and consequent productivity gains.

Effectively, they also are non rival in use by competing plant breeders because they are largely knowledge based. sWhere knowledge is embedded in a tangible technology, the use of which requires consumables that are rival in use, the knowledge component of such technology is still non-rival in use. In effect, the knowledge component enables a capacity to practice the technology that is unlimited, and hence non-rival in use. Hence they share at least one key attribute of a public good, in so far as use by any one plant breeder does not diminish the value of the input to other plant breeders. This capacity to practice key enabling technologies will be referred to as essential plant breeding infrastructure (EPBI). Traditionally, such inputs were non-proprietary, provision was publicly funded, and access by public plant breeding programmes was free and open.

In return, no attempt was made to recover the costs of the breeding programme (as distinct from costs of seed multiplication) by charging growers for the intellectual property embodied in newly released varieties. In a world of privatised plant breeding, these admirable arrangements are unlikely to survive. Current funding sources for the provision and further development of key enabling technologies already are under threat. As a result, public agencies may abandon such activity to the private sector, and/or may seek to recover some or all of the cost by charging plant breeders for access to these technologies. This is likely to hasten an emerging trend towards greater application of intellectual property rights to breeding technologies as well as to germplasm, and to the commercialisation and possible privatisation of their production.

If key enabling technologies are proprietary or otherwise price excludable, then they belong to a class of goods referred to as excludable public goods. This may provide the necessary incentive for private sector investment to compensate for declining public and/or collective industry funding. If it does not, eventually innovation and consequent returns to private investment in plant breeding are likely to stall as a result. However, the stimulus to inventive activity provided by the patent system in other areas demonstrates that private provision of excludable public goods is feasible, and may have advantages vis-Å-vis public provision of pure public goods.

Nonetheless, there also are threats to the efficient provision and utilisation of EPBI that need to be addressed if the potential benefits from some form of privatised provision of key enabling technologies are to be fully realised. For instance, if private firms respond individually rather than collectively, there will be wasteful duplication of effort in producing such inputs as commercial plant breeders strive for competitive advantage in the market place. Moreover, there is the risk that the incentives for an individual private investment will be inadequate for it to substitute fully for the likely withdrawal of public funding. Either of these two scenarios is inefficient. The obvious solution is for cooperative behaviour to provide such inputs. This may involve joint funding by private plant breeders, such as international consortia formed to develop molecular marker technology.

However, given the history of funding of plant breeding, the more likely scenario for most EPBI is a continuation of collective funding by industry through a body such as the GRDC. Note that while "membership" of GRDC is compulsory for all producers of mandate grains, the cost of membership is linked to the level of grain production. Nevertheless, there is increasing pressure for GRDC to commercialise those enabling technologies in which it invests, and to recoup at least part of the cost of its investment by way of user fees. Irrespective of whether funding for future provision of essential plant breeding infrastructure comes for collective funding or private investment, it is highly likely that there will be a lack of competition in the provision of essential plant breeding infrastructure. While adequate and efficient provision of EPBI is one cause for concern, efficient utilisation is another. These inputs are quasi public goods in the sense that they are non-rival in use. As a result, efficient utilisation involves both the potential for under-utilisation of essential plant breeding infrastructure due to monopoly pricing, and the much maligned concept of the "level playing field" in downstream industries.

Economists' interest in this concept stems from the observation that if the institutional, policy, or legal framework confers advantages on some firms relative to others, competition may not generate desirable outcomes. If the favoured firms are not the most efficient, the outcome may be that inefficient producers out competing their more efficient counterparts. Conversely, if all firms compete on a "level playing field", then the law of the jungle should ensure that only the most efficient survive.

There are obvious parallels here to National Competition Policy (NCP) principles governing access to essential infrastructure. The aim of National Competition Policy (NCP) is to facilitate effective competition where competition between suppliers of goods and services result in lower prices, a wider range of products, and/or better service for consumers, but also to accommodate situations where competition does not have that effect, or where it conflicts with social objectives. In industries such as telecommunications, air and rail transport, and electricity transmission, NCP recognises that competition may not be feasible or desirable in the provision of some essential infrastructure, and that the shared use of such 'bottleneck' or 'essential' infrastructure facilities may be necessary to facilitate efficient competition in downstream markets that use such infrastructure.

Access regulation that aims to promote competition in markets that use the services of 'essential' infrastructure while preserving incentives to develop and maintain those facilities have been developed to address concerns about denial of access and/or monopoly pricing of access. A case can be made that, as plant breeding becomes increasingly privatised, equivalent access regimes will need to be developed for essential plant breeding infrastructure (EPBI).

Unless such an access regime is established, some of the potential benefits from scientific discoveries underpinning modern plant breeding may not be fully realised. In common with NCP access regimes, the aim should be to promote full and efficient competition between plant breeders, while preserving adequate incentives for investment in the ongoing development, maintenance, and provision of essential plant breeding infrastructure. GRDC, as the key provider of EPBI for crop breeding, is cognisant of this problem, and is working to develop policies for access by all plant breeders to essential plant breeding infrastructure in which it invests.

A key issue that will need to be addressed in these policies is the grounds (if any) for denial of access or for discriminatory pricing. For example, one possibility would be to deny access to, or charge higher prices for EPBI to large multi-national "life science" firms. A possible ground for doing so would be that these multi-national firms have access to other sources of EPBI from which Australian plant breeding firms are excluded. Another possible ground would be that GRDC has invested in selected plant breeding firms. Fears have been expressed that they may decide to "protect" such investments by limiting other plant breeders' access to GRDC funded EPBI. *Prima facie*, denying access or discriminatory pricing for this reason would seem to be an example of exploiting market power in order to benefit owned or related entities in upstream or downstream markets. Specifically, it would inhibit rigorous competition in the downstream plant breeding market. Nevertheless, there may be grounds for treating overseas owned plant breeding firms differently to domestically owned firms because of the potential impact on Australia's trading position.

DEVELOPMENT OF PLANT BREEDING: PLANT BREEDER

The aim of every plant breeder is to produce a group that will be worth a place in the assortment of cultivated plants. In some cases the new plant will have only ornamental value, in others it will, for any of a dozen reasons, be better adapted to agriculture than other similar ones. It is easy enough in all classes of plants to produce novelties, to produce variability.The experimental grounds of the plant breeder would soon be cluttered up with novelties if he did not keep on weeding out the comparatively worthless.

After producing our novel plants, we must judge them. Some qualities are much easier to judge than others. It is comparatively easy to weed out a lot of seedling dahlias and to pick out a few that may be worth keeping on for further testing. In the selection of our agricultural plants, in cereals, in beets, in maize, the assortment of valuable kinds is already so great that only relatively few of the newer selections are worth keeping. In the vegetatively propagated plants this work of testing novelties is comparatively easy.

In the first place, we are not bothered by considerations of purity—once we have produced a very good specimen, we can just multiply it with all its properties intact. In the second place, we can easily, for purposes of testing and comparison, produce as many individual plants of the novelty as we require. When the differences are relatively small—as in cereals and beets, where the novel plant produced must compete with excellent old stock that has been made by a process of selection and comparison during centuries—a correct system of evaluation becomes of very great importance. What we really want to test is the inherited composition of the material. This may be difficult, as the final quality of the plants to a large extent depends on favourable or unfavourable conditions under which they grew up. It will perhaps be as well to begin with the simplest problems. In the production of novelties in such flowers as roses or dahlias the value of a new variety depends entirely on the appeal it makes to the gardening public. The man who is proud to have found a novel dahlia seedling or a new and strikingly beautiful rose sends it to a show and hopes that it will take the judge's eye and win him the much- "prized ribbon or certificate. But of course the insider knows that many highly commended novelties will never be heard of again, or that they will drop out of cultivation in a few years. There is a difference between the fact that the grower can bring half a dozen perfect blooms to the show on time, and the question whether that same rose or chrysanthemum will ever be popular in the gardens, or even be worth while growing for the commercial producer of cut flowers. A dahlia that wins the medal may be a very weak grower, or it may be very susceptible to rain.

The strikingly beautiful rose may be very susceptible to mildew, or its flowering season may be extremely short. If we visit the growers of greenhouse roses, or the gardens of the man who produces cut flowers for the market, we find that the number of varieties that have real commercial value—varieties "with bread in them ", as our gardeners express it—is extremely small. We shall probably find only three or four of the relatively older dahlias worth growing for cut flowers, and it is seldom that a novelty is added to that list.

In testing out agricultural plants, both during the process of selection and after they have finally been introduced, many methods are in use. The direct, simple examination of a number of small plots or rows will show us the qualities of those lots, but this method will only serve to find the groups with definitely inferior quality. It is not good enough to find exceptional merit, as circumstances may have favoured those qualities we appreciate. The only way to avoid reaching the wrong conclusions is to see to it that environmental conditions do not vitiate our results. The best way to do this is to com- pare the quality of a number of plants with that of other plants that were grown in wholly comparable conditions.

One of the best ways to do this is to grow the plots of our plant between similar plots of an older variety, that can be used as a standard of comparison. The choice of this standard is of great importance, and we will revert to this subject when treating of the breeding of special plant groups. As an example we can here take the evaluation of a newly produced (or of an imported) potato clone, or pure line of wheat or barley, In a case like this the standard may be some line or clone that is of commercial importance, against which the tested plant will be called upon to compete. If there is sufficient material, it is wise to repeat the comparison—that is to say, we should grow the test plot between standard plots in different parts of the field, or preferably in different fields. To equalize accidental conditions in small parts of the field, oblong plots are generally better than square ones. If we are testing a number of experimental lots, we can grow them side by side, and interpolate standard plots to be used for purposes of comparison.

The experimental garden will need an efficient system of book-keeping. We must note down anything of value about the experimental and the standard plots, date of maturity, disease resistance, quality, yield. We can calculate average yield for the sum of the plots of one kind, and then compare those figures with the average computed yield of the standard plots or with the average yield of the whole field; and a comparison of the deviation of the yields in the different plots with calculated mean errors will give us valuable data about the question in how far the superiority or inferiority of each separate number is significant as an indication of merit.

The practice of plant breeding in different parts of the world. A multitude of miserably small checkerboard fields being weighed at harvest time to produce the figures for an office staff to juggle with. A successful establishments where the field-work was under the supervision of men who were thoroughly conversant with the material, and where excellent results were obtained with a minimum of calculating machine work. A method of annotating the results obtained in experimental plots which gives the plant breeder an excellent insight into the merits of his new material is that of graphical registration. This method can be adapted to the most diverse projects; it is very simple, and gives us our results at a glance. The method consists of registering the quality (yield, height, resistance) of each plot, both standard plots and testing plots, on ruled paper.

If we do this, the line that can be drawn through the points of the standard plots will probably rise in one part of the field, and descend in some other part of it, and by drawing a line that joins those points we obtain a basis upon which we can found our judgement of the results obtained in the experimental plots. If we want to test our experimental material in two or three different qualities, we must construct separate graphs for each of those qualities, or sometimes we can use the same graph, employing different colours for separate qualities (weight, sugar percentage, purity in beets, etc.).

Reading those graphs will now show us the quality of our plant plots, in comparison with the standards, and we can use this relative value as a basis for our selection work. If we compare half a dozen groups in one experiment, and we find that one or two of them will in every duplication remain well below the standard in quality, this may be a very good reason for discarding them. On the other hand, if we find one or two numbers that all through the experiment are well above the standard line in quality, we have a very good reason for preferring them. It may be that in the case of one or two numbers the results are very variable, so that in one part of the field the number is above and in other parts it is below the standard in quality. Then we may well retain this number for another trial in the next season.

The influence of one row of plants on those growing close to it may be considerable, and we must guard against this. If we can grow three rows and use only the middle one for our records, we take a step in the right direction. If we have enough seed to plant six rows, and we examine only the two middle ones, discarding two rows on each side, this is still better. The greatest difficulty arises in those cases where we are dealing with very few plants of each number; or, worse, where, as in the second generation of a cross, every plant is probably different from every other. It then becomes very important to equalize conditions for each plant as far as this is possible. A very good plan is to grow the plants in rows, and alternate those experimental rows with rows of some pure standard variety. Each plant may then have two diverse neighbours in its own row, but it will touch four other plants that will be of the same kind as four out of six neighbour plants in the case of every other plant in the experiment.

This means that to save time we must make a preliminary choice out of the many numbers in those early generations, and it is very important that we shall make the cultural conditions as nearly equal as possible for every plant. This is the place to warn against using apparent correlative qualities in judging the value of our experimental plants. In the breeding of animals even to-day we often see that almost as much attention is paid to correlative qualities as to real merit.

Some of this is certainly due to the fact that in some material real merit can only be judged in one of the two sexes. There is no good reason to give any attention to correlative qualities. In so far as a real correlation exists, selection for real quality will automatically bring along the correlative character, and in as far as we are mistaken (and this will nearly always be so) we are simply led astray by paying attention to apparent correlations. In plant-breeding selection according to qualities believed to be indirect indications of good quality used to be very frequently attempted.

This has gone out of fashion, and we only meet traces of it in very special fields (sugar-cane, apples). In our selection work we should always consider actual practical merit, and we should care- fully study in how far the qualities we can examine, measure and weigh are indications of this real merit.

BREEDING METHODS IN CROP PLANTS

SELF POLLINATED CROPS

Mass selection: In mass selection, seeds are collected from (usually a few dozen to a few hundred) desirable appearing individuals in a population, and the next generation is sown from the stock of mixed seed. This procedure, sometimes referred to as phenotypic selection, is based on how each individual looks. Mass selection has been used widely to improve old "land" varieties, varieties that have been passed down from one generation of farmers to the next over long periods.

An alternative approach that has no doubt been practiced for thousands of years is simply to eliminate undesirable types by destroying them in the field. The results are similar whether superior plants are saved or inferior plants are eliminated: seeds of the better plants become the planting stock for the next season.

A modern refinement of mass selection is to harvest the best plants separately and to grow and compare their progenies. The poorer progenies are destroyed and the seeds of the remainder are harvested. It should be noted that selection is now based not solely on the appearance of the parent plants but also on the appearance and performance of their progeny. Progeny selection is usually more effective than phenotypic selection when dealing with quantitative characters of low heritability. It should be noted, however, that progeny testing requires an extra generation; hence gain per cycle of selection must be double that of simple phenotypic selection to achieve the same rate of gain per unit time. Mass selection, with or without progeny test, is perhaps the simplest and least expensive of plant-breeding procedures. It finds wide use in the breeding of certain forage species.

Pure-line selection: Pure-line selection generally involves three more or less distinct steps: (1) numerous superior appearing plants are selected from a genetically variable population; (2) progenies of the individual plant selections are grown and evaluated by simple observation, frequently over a period of several years; and (3) when selection can no longer be made on the basis of observation alone, extensive trials are undertaken, involving careful measurements to determine whether the remaining selections are superior in yielding ability and other aspects of performance.

Any progeny superior to an existing variety is then released as a new "pure-line" variety. Much of the success of this method during the early 1900s depended on the existence of genetically variable land varieties that were waiting to be exploited. They provided a rich source of superior pure-line varieties, some of which are still represented among commercial varieties. In recent years the pure-line method as outlined above has decreased in importance in the breeding of major cultivated species; however, the method is still widely used with the less important species that have not yet been heavily selected. A variation of the pure-line selection method that dates back centuries is the selection of single-chance variants, mutations or "sports" in the original variety. A very large number of varieties that differ from the original strain in characteristics such as colour, lack of thorns or barbs, dwarfness, and disease resistance have originated in this fashion.

Hybridization

During the 20th century planned hybridization between carefully selected parents has become dominant in the breeding of self-pollinated species. The object of hybridization is to combine desirable genes found in two or more different varieties and to produce pure-breeding progeny superior in many respects to the parental types.

Genes, however, are always in the company of other genes in a collection called a genotype. The plant breeder's problem is largely one of efficiently managing the enormous numbers of genotypes that occur in the generations following hybridization. As an example of the power of hybridization in creating variability, a cross between hypothetical wheat varieties differing by only 21 genes is capable of producing more than 10,000,000,000 different genotypes in the second generation. At spacing normally used by farmers, more than 50,000,000 acres would be required to grow a population large enough to permit every genotype to occur in its expected frequency. While the great majority of these second generation genotypes are hybrid (heterozygous) for one or more traits, it is statistically possible that 2,097,152 different pure-breeding (homozygous) genotypes can occur, each potentially a new pure-line variety. These numbers illustrate the importance of efficient techniques in managing hybrid populations, for which purpose the pedigree procedure is most widely used.

Pedigree breeding starts with the crossing of two genotypes, each of which have one or more desirable characters lacked by the other. If the two original parents do not provide all of the desired characters, a third parent can be included by crossing it to one of the hybrid progeny of the first generation.

In the pedigree method superior types are selected in successive generations, and a record is maintained of parent–progeny relationships. The F2 generation (progeny of the crossing of two F1 individuals) affords the first opportunity for selection in pedigree programmes. In this generation the emphasis is on the elimination of individuals carrying undesirable major genes.

In the succeeding generations the hybrid condition gives way to pure breeding as a result of natural self-pollination, and families derived from different F2 plants begin to display their unique character. Usually one or two superior plants are selected within each superior family in these generations. By the F5 generation the pure-breeding condition (homozygosity) is extensive, and emphasis shifts almost entirely to selection between families.

The pedigree record is useful in making these eliminations. At this stage each selected family is usually harvested in mass to obtain the larger amounts of seed needed to evaluate families for quantitative characters. This evaluation is usually carried out in plots grown under conditions that simulate commercial planting practice as closely as possible.

When the number of families has been reduced to manageable proportions by visual selection, usually by the F7 or F8 generation, precise evaluation for performance and quality begins. The final evaluation of promising strains involves:

- Observation, usually in a number of years and locations, to detect weaknesses that may not have appeared previously;
- Precise yield testing; and
- Quality testing.

Many plant breeders test for five years at five representative locations before releasing a new variety for commercial production. The bulk-population method of breeding differs from the pedigree method primarily in the handling of generations following hybridization. The F2 generation is sown at normal commercial planting rates in a large plot. At maturity the crop is harvested in mass, and the seeds are used to establish the next generation in a similar plot. No record of ancestry is kept. During the period of bulk propagation natural selection tends to eliminate plants having poor survival value. Two types of artificial selection also are often applied:

- Destruction of plants that carry undesirable major genes and
- Mass techniques such as harvesting when only part of the seeds are mature to select for early maturing plants or the use of screens to select for increased seed size.

Single plant selections are then made and evaluated in the same way as in the pedigree method of breeding. The chief advantage of the bulk population method is that it allows the breeder to handle very large numbers of individuals inexpensively.

Often an outstanding variety can be improved by transferring to it some specific desirable character that it lacks. This can be accomplished by first crossing a plant of the superior variety to a plant of the donor variety, which carries the trait in question, and then mating the progeny back to a plant having the genotype of the superior parent.

This process is called backcrossing. After five or six backcrosses the progeny will be hybrid for the character being transferred but like the superior parent for all other genes. Selfing the last backcross generation, coupled with selection, will give some progeny pure breeding for the genes being transferred. The advantages of the backcross method are its rapidity, the small number of plants required, and the predictability of the outcome.

A serious disadvantage is that the procedure diminishes the occurrence of chance combinations of genes, which sometimes leads to striking improvements in performance.

Hybrid Varieties

The development of hybrid varieties differs from hybridization. The F1 hybrid of crosses between different genotypes is often much more vigorous than its parents. This hybrid vigour, or heterosis, can be manifested in many ways, including increased rate of growth, greater uniformity, earlier flowering, and increased yield, the last being of greatest importance in agriculture.

CHOICE OF PLANT-BREEDING MATERIAL

The advantage of having at one's disposal a large collection of plant varieties, both plants of commercial importance and imported plants of no immediate practical importance, is enormous. A difficult question for the plant breeder is always what crosses to make, what plants to combine with the greatest hope of ultimate success. In the breeding of such plants as potatoes, raspberries and roses, certain clones have the reputation of being profitable parents of hybrid lots. The Early Rose potato, the Lloyd George raspberry, the Chilean strawberry. In some special instances the choice is obvious. When our present assortment of flaxes or potatoes or roses happens to be deficient in some such quality as disease resistance or winter- hardiness, we are on the look-out for plants that will show those desirable qualities to the greatest extent. In other words, we shall, when choosing the parents of our future hybrids, try to find material that promises the possibility of a combination of the very good points of both.

This is obvious. Often crosses that seemed very little promising have given excellent results. It would seem that here no definite rule could be indicated. It is always worth while trying crosses with new material, with newly imported species—in fact with material which has not been used previously. This is especially so when, as in horti- culture, everybody is on the look-out for something new. Another rule that could be given would be that when we make very wide crosses we cannot hope to find commercially valuable strains in the second hybrid generation, unless we grow extremely large numbers of plants in that generation. When we make a very wide cross to incorporate some special quality into the existent stocks, the best thing we can do is to use the hybrids in back-cross series.

We can then strive to retain just one special quality of the new species and incorporate it into the general set of genes with which we are familiar in our older stock. In several instances very wide crosses have been successfully used, and sometimes plants have been used in cross-breeding that in themselves showed very little promise. The inedible cattle gourd that Orton crossed with his water- melons, of some of the potato crosses with Andean wild species (Salaman) and of the back-crosses to hybrids with wild cane from which all our modern resistant varieties of sugar-cane have been derived.

It is here—in the choosing of plant material for breeding work—that cytological studies and chromosomal counts are of the greatest practical value.

CROSS POLLINATED CROPS

The most important methods of breeding cross-pollinated species are:

- Mass selection;
- Development of hybrid varieties; and
- Development of synthetic varieties.

Since cross-pollinated species are naturally hybrid (heterozygous) for many traits and lose vigour as they become purebred (homozygous), a goal of each of these breeding methods is to preserve or restore heterozygosity.

Mass Selection

Mass selection in cross-pollinated species takes the same form as in self-pollinated species; *i.e.,* a large number of superior appearing plants are selected and harvested in bulk and the seed used to produce the next generation. Mass selection has proved to be very effective in improving qualitative characters, and, applied over many generations, it is also capable of improving quantitative characters, including yield, despite the low heritability of such characters. Mass selection has long been a major method of breeding cross-pollinated species, especially in the economically less important species.

Hybrid Varieties

The outstanding example of the exploitation of hybrid vigour through the use of F1 hybrid varieties has been with corn (maize). The production of a hybrid corn variety involves three steps:

- The selection of superior plants;
- Selfing for several generations to produce a series of inbred lines,

which although different from each other are each pure-breeding and highly uniform; and

- Crossing selected inbred lines.

During the inbreeding process the vigour of the lines decreases drastically, usually to less than half that of field-pollinated varieties. Vigour is restored, however, when any two unrelated inbred lines are crossed, and in some cases the F1 hybrids between inbred lines are much superior to open-pollinated varieties. An important consequence of the homozygosity of the inbred lines is that the hybrid between any two inbreds will always be the same. Once the inbreds that give the best hybrids have been identified, any desired amount of hybrid seed can be produced.

Pollination in corn (maize) is by wind, which blows pollen from the tassels to the styles (silks) that protrude from the tops of the ears. Thus controlled cross-pollination on a field scale can be accomplished economically by interplanting two or three rows of the seed parent inbred with one row of the pollinator inbred and detasselling the former before it sheds pollen. In practice most hybrid corn is produced from "double crosses," in which four inbred lines are first crossed in pairs (A × B and C × D) and then the two F1 hybrids are crossed again (A × B) × (C × D). The double-cross procedure has the advantage that the commercial F1 seed is produced on the highly productive single cross A × B rather than on a poor-yielding inbred, thus reducing seed costs. In recent years cytoplasmic male sterility, described earlier, has been used to eliminate detasselling of the seed parent, thus providing further economies in producing hybrid seed. Much of the hybrid vigour exhibited by F1 hybrid varieties is lost in the next generation. Consequently, seed from hybrid varieties is not used for planting stock but the farmer purchases new seed each year from seed companies.

Perhaps no other development in the biological sciences has had greater impact on increasing the quantity of food supplies available to the world's population than has the development of hybrid corn (maize). Hybrid varieties in other crops, made possible through the use of male sterility, have also been dramatically successful and it seems likely that use of hybrid varieties will continue to expand in the future.

Synthetic Varieties

A synthetic variety is developed by intercrossing a number of genotypes of known superior combining ability—*i.e.,* genotypes that are known to give superior hybrid performance when crossed in all combinations. (By contrast, a variety developed by mass selection is made up of genotypes bulked together without having undergone preliminary testing to determine their performance in hybrid combination.) Synthetic varieties are known for their hybrid vigour and for their ability to produce usable seed for succeeding seasons. Because of these advantages, synthetic varieties have become increasingly favoured in the growing of many species, such as the forage crops, in which expense prohibits the development or use of hybrid varieties.

BREEDING PERENNIAL GRAINS: WARM-SEASON GRASSES

SORGHUM

Hybridization with Sorghum propinquum: In tropical environments, grain sorghum (*S. bicolor,* 2n=20) is able to regrow from basal nodes to produce a rattoon crop. But breeding a sorghum that is winterhardy in temperate regions will require transfer of genes from related species.

A perennial native of southeast Asia, *S. propinquum* is rhizomatous and diploid, with chromosomes largely homologous to those of grain sorghum. Paterson *et al.,* evaluated rhizome-related traits of 370 F2 and 378 BC1 plants from a cross between the two species. Surviving a mild winter in southern Texas, USA, with only three nights reaching temperatures of -3°C to -4°C, 92% of F2 plants and 46% of BC1 regrew in the spring. Plants regrew either from tillers or from rhizomes. Forty-eight F2 plants representing the range of the population were selected for progeny testing. From all F2 plants and F3 lines, Paterson *et al.,* collected data on number of rhizomes producing above-ground shoots, distance between the centre of the crown and the most distal shoot, a subterranean rhizome score, tillering, and regrowth. They mapped chromosomal segments affecting these traits in the F2 plants and F3 lines, using 78 RFLP loci.

Rhizomatousness was a complex trait, with nine different chromosomal regions on seven of sorghum's ten chromosomes having detectable effects on at least one of the rhizome traits. Individual segments accounted for between 5 and 13% of the total variation. All but one of the seven segments associated with regrowth was also associated with one or more rhizome traits, and all four segments associated with tillering were also associated with regrowth or rhizomatousness.

Because *S. propinquum* is a tropical species, rhizomatous progeny of crosses between *S. bicolor* and *S. propinquum* would probably not be winterhardy at middle or northern latitudes, without successful selection for deeper rhizome growth. But there is great potential for developing a perennial grain sorghum for the tropics or subtropics from such populations.

Hybridization with johnsongrass: Johnsongrass (*S. halapense,* 2n=40) is a tetraploid, probably an amphiploid that combines the genomes of *S. bicolor* and *S. propinquum*. It is a very strong and aggressive perennial, and a notorious weed. Like *S. propinquum,* johnsongrass stores starch in its rhizomes. As a consequence, its rhizomes have no cold hardiness, unlike those of temperate grasses, which store fructosans. Natural selection for deeper-growing rhizomes has allowed johnsongrass to spread as a weed as far north as Ontario. Although the most northerly biotype reproduces mainly by seed, regrowth from rhizomes occurs throughout the range of the species.

Early research on hybrids between diploid sorghum and johnsongrass produced two types of hybrids: 30-chromosome plants that were male-sterile but could be backcrossed to the diploid parent and fertile 40-chromosome plants derived from unreduced female gametes in the diploid parent. The 30-chromosome plants were more strongly rhizomatous. Hadley and Mahan identified seven 20-chromosome backcross plants that were rhizomatous, but most were chlorophyll mutants. Three years of selection failed to produce a single diploid line that was rhizomatous.

Hybrids between *S. halapense* and induced tetraploid lines of *S. bicolor* are easily made. In an effort to produce a perennial grain sorghum at The Land Institute, Piper and Kulakow crossed *S. halapense* with tetraploid grain sorghum lines. In an interspecific F3 population, approximately 40% of plants were rhizomatous. There was no significant negative correlation between rhizome production and grain yield in the F3 generation, but a negative association arose with backcrossing. Yield was strongly related to plant biomass and root biomass, both with pheontypic correlations of 0.70. The F3 population — selected for winterhardiness but not for yield — had a grain yield 62% as high as the mean non-irrigated sorghum yield in Saline County, Kansas, where the experiments were conducted. But rhizome production dropped to near zero in other populations derived by backcrossing to tetraploid *S. bicolor* in an effort to increase grain yield.

Land Institute breeders have selected for winter survival among the rare rhizomatous BC2 plants and their selfed progeny. The phenotypes of winterhardy selections remain very distant from that of the cultivated parent, despite the latter's expected 87.5% genetic contribution to the BC2. The selections are taller and later maturing than either parent, have open panicles and small seed, and produce many tillers, although not as many as johnsongrass. Rhizome mass is less than 10% that of johnsongrass — sufficient for overwintering, but not enough to allow interspecific progenies to become aggressive weeds.

Both rhizomes and tillers originate from meristems at the base of the plant, and there appears to be considerable overlap in their genetic control. It may not be possible, or even desirable, to select a low-tillering, sufficiently rhizomatous genotype. Indeed, selection for yield improvement may be more effective in highly tillering populations. If a grass plant is regarded as a population of largely autotrophic tillers, and each tiller supports both seeds and rhizomes, then the most direct route to increased yield is via additional tillers. Of course, this implies increased biomass.

Piper and Kulakow concluded that development of a winterhardy sorghum (*i.e.*, one that produces 80g of rhizomes per plant) with a grain yield of over 4000 kg/ha is feasible, through selection for greater biomass and reallocation of photosynthate to seed production. The foundation germplasm for breeding a perennial sorghum may necessarily consist of high-biomass plants that produce more tillers than annual sorghum. With a sufficiently large genetic base, subsequent selection for improved harvest index and seed size could be successful. Because there is some homology between the chromosomes of grain sorghum and johnsongrass, multivalent chromosome associations are common at meiosis in interspecific tetraploids. Multivalents, in turn, cause poor seed set because of nondisjunction of chromosomes. Luo *et al.,* demonstrated that selection for fertility can be effective in autotetraploid grain sorghum, which is generally plagued by low seed set. Breeders could cross perennials with the highly fertile tetraploid germplasm that Luo *et al.,* have produced. Broadening and improving the genetic base of tetraploid perennial sorghum will require introduction of more agronomically elite germplasm. One rapid method of incorporation would be to pollinate both diploid and induced-tetraploid strains of elite, large-seeded inbred lines with the best tetraploid perennials. From the diploid/tetraploid crosses, breeders can select 40-chromosome hybrids that arise from unreduced gametes. The Land Institute is now taking this approach to develop genetically diverse breeding populations.

Pearl Millet

Pearl millet (*Pennisetum glaucum,* 2n=14), like sorghum, is a tropical, annual diploid with a perennial, tetraploid relative. Napiergrass, *P. purpureum* (2n=28) has one genome homologous and one nonhomologous to that of pearl millet. Amphiploids resulting from colchicine treatment of hybrids between the species are male and female fertile. Hanna backcrossed these hexaploids to diploid and tetraploid pearl millet lines and produced perennial progeny; however, both types of backcross plants (tetraploid and pentaploid) were highly sterile and unable to survive the mild winters of south Georgia, USA. Napiergrass, the only known species in pearl millet's secondary gene pool, is not rhizomatous, so selection for winterhardiness would probably not be successful. A perennial millet for grain production in the tropics is a reasonable prospect.

Dujardin and Hanna interpollinated hybrids and their derivatives from crosses between tetraploid pearl millet and *P. squamulatum* (2n=54), a more distant, perennial, apomictic relative. Some progenies (2n=48) were both perennial and apomictic. Apomixis can be used to ensure grain production and genetic stability in highly heterozygous progenies of interspecific crosses. *Pennisetum squamulatum* is also non-rhizomatous, and its progeny are not likely to be perennial outside of the tropics.

Maize

Hybridization with tetraploid perennial teosinte: Efforts to develop perennial maize (*Zea mays* ssp. *mays,* 2n=20) have been sporadic at best; as in other crops, hybridization between maize and perennial relatives has led primarily to improvement of the annual crop. Shaver first attempted development of maize-like perennials from crosses between colchicine-induced tetraploids of maize and a wild, perennial, tetraploid relative, *Z. mays* ssp. *perennis* (2n=40). Selection within the resulting tetraploid populations and backcrosses to tetraploid maize effectively increased the frequency of perennial progeny. Crosses to diploid maize produced perennial triploids, but all diploid selections were annual.

Shaver combined a postulated gene (*pe*) for perenniality with recessive genes for indeterminacy (*id*) and grassy tillers (*gt*) in a diploid background, to produce perennial plants; however, the *idid* genotype prevented production of ears. Because Shaver had developed a separate *idid* population in a different genetic background that did produce ears, he suggested that perennial diploids could also be made fertile if the genetic background were manipulated.

Hybridization with diploid perennial teosinte: Little further attention was paid to perennial maize until the dramatic discovery of a diploid species of perennial teosinte, *Z. mays* ssp. *diploperennis*.

Initial studies showed that inheritance of perenniality was relatively simple in maize/*diploperennis* crosses, but perenniality was inferred from tillering habit, a potentially misleading technique. In subsequent, larger-scale experiments, inheritance of tillering in progeny of similar inter-subspecific crosses was more complex, and perennial maize types were not recovered even in large segregating populations.

Genetic mapping in maize/annual teosinte crosses show that most traits of domestication separating the species are oligogenic, and the loci tend to be clustered on the map, through either linkage or pleiotropy. A similar study of these traits, plus perenniality, in crosses between maize and diploid perennial teosinte would be of great value to any breeding programme attempting to combine perenniality with the agronomic phenotype of maize. This would require substantial effort to evaluate large segregating populations for tillering, rhizome production, and capacity to produce seed over multiple seasons. Once the genomic regions of interest are identified, marker-assisted selection can be used to incorporate them into a maize background and eliminate unwanted alleles such as those conditioning hard glumes and shattering.

One serious obstacle to adoption of any teosinte-derived perennial grains is the lack of winterhardiness of these tropical species. There are no winterhardy species of *Zea.* Because the bulk of maize production and breeding occurs in temperate areas, there has been little incentive to develop perennials from crosses with *Z. mays* ssp. *diploperennis.*

One possible approach has not been suggested to date: selection for rhizome depth. As we have seen, johnsongrass rhizomes also are not winterhardy if near the soil surface, but dispersal of the species into higher latitudes has been made possible by selection for deeper rhizomes. Superimposing selection for this undoubtedly complex trait on selection for perenniality and traits of domestication, not to mention yield, may entail a much larger effort than any breeding programme is willing to undertake.

Hybridization with eastern gamagrass: The closest winterhardy relatives of maize are in the genus *Tripsacum.* Eastern gamagrass (*T. dactyloides*), for example, is currently grown as a perennial forage grass as far north in the western hemisphere as Kansas and Massachusetts, and can be grown in the Corn Belt. *T. dactyloides* has been hybridized many times with maize, beginning with the work of Manglesdorf and Reeves. Plants of the diploid (2n=36) or tetraploid (2n=72) races may be crossed with maize. If *Tripsacum* is used to pollinate maize, embryo rescue is necessary, but if maize is used as the male, some hybrid seed may be obtained without rescue. In addition, several strains of popcorn, when pollinated with tetraploid *T. dactyloides,* produce large amounts of hybrid seed that does not require embryo rescue. Some have good crossability with diploid *T. dactyloides* as well. Contrary to typical results, Eubanks reported that a putative 20-chromosome hybrid between *T. dactyloides* and *Z. diploperennis* showed 93 to 98% pollen fertility.

Natural introgression between *Tripsacum* and maize has not been observed, but morphological and molecular evidence supports the hypothesis that the species *T. andersonni* is an intergeneric hybrid containing three genomes (54 chromosomes) from *Tripsacum* and 10 chromosomes from *Zea* in *Tripsacum* cytoplasm. The uniformity of this ancient natural hybrid indicates that *T. andersonii* arose from a single hybridization. It has been able to spread across tropical Latin America because of its vigorous perenniality.

In addition to being perennial, tetraploid *T. dactyloides* is a facultative apomict. Perennial hybrids result from artificial crosses between tetraploid *Tripsacum* and maize, and some seed-set can result from apomixis. The hybrids, derived from parents with different basic chromosome numbers and chromosomes of different sizes (those of maize being larger), are male sterile, with cytological behaviour that is anything but regular.

Harlan and deWet summarized methods for utilizing such hybrids in maize improvement. Either 28-chromosome or 46-chromosome hybrids — derived from diploid and tetraploid *T. dactyloides* parents, respectively — can be backcrossed to maize. In either case, *Tripsacum* chromosomes are eliminated with backcrossing. Elimination occurs more gradually in progeny of 46-chromosome hybrids, and the 20 chromosomes of the resulting backcross plants can contain significant genetic material from *Tripsacum.*

All 20-chromosome backcross plants derived to date have been annual and non-apomictic. Kindiger *et al.,* derived an annual, 39-chromosome line that carried 9 *Tripsacum* chromosomes and displayed an intermediate level of apomixis. Most hybridization with *Tripsacum* has been for the purpose of either elucidating the evolution of maize or transferring resistance or other genes to annual maize. The latter purpose implies backcrossing to maize. But development of perennial populations may require interpollinating plants in early backcross generations that still carry many *Tripsacum* chromosomes, or even backcrossing to *Tripsacum.* As Harlan and deWet commented, "Apparently, if one wishes to contaminate maize with *Tripsacum* one should first contaminate *Tripsacum* with maize."

New approaches to perennial maize are being explored. An anomalous fertile hybrid between diploid *T. dactyloides* and maize was discovered by one of the authors (BEZ) in 1997 near the mouth of the Big Nemaha river in Richardson County, Nebraska, USA. This derivative of natural introgression between *T. dactyloides* and a putative commercial hybrid is being hybridized with gynomonoecious *Tripsacum* — both diploid and tetraploid — and with tassel-seed popcorn to develop a 56-chromosome perennial cultivar for production of grain, forage, fibre, and fuel.

The difficulties encountered in introgressing apomixis from *Tripsacum* into maize should temper hopes for a rapid synthesis of perenniality with high grain yield. Whatever the initial population, and even with marker-assisted selection, the process of recovering perennial, winterhardy segregants with maize-like ears, and then breeding for yield and other agronomic traits will be long and arduous.

Direct Domestication of Warm-season Grasses

Eastern gamagrass: Could *T. dactyloides* be domesticated directly, without introgression of genes from maize? To do so would be an accomplishment parallel to that of domesticating maize from annual teosinte — a feat requiring thousands of years and producing genetic and physiological changes much greater than those involved in domestication of Asian cereals such as wheat and rice. To develop a crop from eastern gamagrass using the knowledge and techniques provided by 21st-century genetics, while leaving aside the important genes of domestication available in maize, would be an ambitious project.

Wagoner described in detail the status of eastern gamagrass as a potential grain crop, and the species' most discouraging characteristic: very low seed yield. Interest had been stimulated by the discovery of a gynomonoecious, or pistillate, mutant in which pistillate and perfect spikelets replace the staminate spikelets of the normal inflorescence. The result is an increase of up to 20-fold in the number of seeds produced per plant; however, the seeds are small, so that the weight of seed produced per plant is increased by only a factor of 3. Plants of eastern gamagrass are large, vigorous and widely adapted. The increase in sink size made possible by the pistillate mutant may provide an opportunity to increase seed yield dramatically through increased harvest index— the yield component usually found to have had the greatest effect on yield improvement in traditional grain crops. Furthermore, Jackson and Dewald found that the increased seed yield of pistillate plants did not come at the expense of plant vigour or longevity. Carbohydrate reserves were significantly higher in pistillate than in normal genotypes. For breeders, there is a huge pool of genetic variability available in the species. Although tetraploid *T. dactyloides* reproduces apomictically, parental combinations can be produced via BIII hybrids, in which an unreduced egg is fertilized by a haploid sperm. Alternatively, obligately sexual tetraploid plants can be produced via colchcine treatment of diploid *T. dactyloides* plants, all of which are sexual. The genus *Tripsacum* contains many species that lack winterhardiness but have desirable traits that, potentially, could be transferred to eastern gamagrass: synchronous flowering, large spikelet number, higher seed yield, and other variations in plant morphology.

The maximum seed yield of eastern gamagrass in plots at The Land Institute has been 240 kg/ha, in the third year after sowing. It remains to be seen how rapidly yield can be improved through selection within pistillate populations. And there is another question: would the 10-, 20-, or 30-fold yield improvements required to make eastern gamagrass a viable grain crop have larger negative effects on plant vigour and persistence than did the three-fold yield boost brought about by the pistillate mutation?

That increase was large relative to the grain yield of a normal plant but required diversion of only a small amount of photosynthate, relative to the plant's large biomass. The food quality of Eastern gamagrass is excellent. But, even if yield can be improved, other problems must be solved. One problem is disease.

Infection by maize dwarf mosaic virus B has been very serious in plots at The Land Institute. Also, the hard fruitcase of *Tripsacum* weighs almost three times as much as the seed itself and makes processing difficult. It was an extremely rare mutation in annual teosinte that freed the kernel from the fruitcase and allowed its use as a grain and the development of maize. This may be a gene that breeders will be forced to transfer from maize.

Indian ricegrass: Grain is currently being harvested from a perennial grass for human food in northeastern Montana, USA. Indian ricegrass (*Oryzopsis hymenoides*), cultivar Rimrock, has reduced seed-shattering and produces gluten-free grain.

The grain is being produced, milled, and marketed under the trade name Montina. Yields vary between 250 and 500 kg/ha, but improvement through breeding may be feasible. Germplasm collections exhibit great phenotypic diversity, and very large-seeded genotypes are known (T.A. Jones, USDA-ARS, Logan Utah, pers. commun.). Certainly, other perennial grasses native to the western USA could be considered for domestication as grain producers, but no attempts have been made.

TECHNOLOGY: TOWARDS AN EVOLUTIONARY PLANT BREEDING

Commercial varieties enter the market through a system of trials and official release. That system requires uniform and stable varieties, and the breeding work must be streamlined for such end results. But local selection is not constrained by those requirements and therefore enjoys some freedom that does not exist within the formal system. Farmer-breeders are free to distribute heterogeneous varieties and can allow crops to continue evolving.

Exploiting heterogeneity and crop evolution in farmers' fields are outside the scope of most plant breeding research. One exceptional experiment, however, has shed some scientific light on the issue. It was started at the University of California (UC) in 1928. Composite cross populations of barley were produced, some of which were extremely diverse in origin of sources. These populations were exposed to continuous natural selection in current modern farming environments and became the subject of studies during the career span of several generations of UC professors. It appears that after low yields in initial years, the composite cross populations gradually improved in performance and eventually became quite good yielders, with excellent yield stability and disease resistance. These results inspired Suneson to propose an evolutionary plant breeding method. After assessment of later generations of the same material, Soliman and Allard concluded that such evolutionary breeding "is unwarranted" if yield potential is the major goal. However, if disease resistance and yield stability are two major objectives, "the composite cross approach is an efficient method". This amounts to saying that a major part of world agriculture, many high-input systems included, could be well served by this approach.

In short, this means constructing a body of broadly diversified germplasm and exposing it to natural selection in areas of contemplated use. For those who are familiar with traditional farmers' breeding, this may sound like reinventing the wheel. In fact it is an improvement of the old wheel of plant breeding. The first step, constructing a body of of broadly diversified germplasm, is not all that straightforward. Science has access to world collections and information sources that are unavailable to farmers. A research institute can chose relevant germplasm and make composite cross populations with an evolutionary potential, most probably far beyond that of locally available varieties.

The immediate outcome, the early generation composite cross population, will be unadapted everywhere and is likely to yield poorly. With time, however, recombinations and natural sorting will improve the adaptation, and, according to the Californian experience, narrow the gap with commercial varieties. The long term outcome could be populations that outperform commercial varieties in disease resistance and yield stability and that may be used as a source of artificial selection for high yield. The disease resistance appears to have evolved through the building up of polygenic complexes. Therefore, it provides a durable resistance as opposed to the monogenic and, therefore, mostly non-durable resistance usually bred into commercial pure line varieties. The stability, at least to some degree, depends on the buffering effect of crop heterogeneity. The Californian experiment shows that when the population is propagated in isolation for a very long time, diversity will start declining, resulting eventually in reduced stability.

Taking the lessons from these experimental findings into the context of current development needs, a few conclusions can be drawn. We need breeding populations with a very high evolutionary potential, and these populations must be exposed to the stress conditions of, or similar to, current farm environments. Furthermore, a certain level of diversity within populations must be maintained in order to sustain evolutionary potential and yield stability. Landraces are usually found to be heterogeneous. But do they have the evolutionary potential required by current development needs? Are they being subjected to a selection pressure that ensures yield stability, and are they as high yielding as they could be? Scientific evidence may not be available to give direct answers to such questions. Community visits may not be helpful either. A confusing picture of different selection practices and frequent change of seeds, appears in many communities where traditional seed systems prevail. It is also common to see the coexistence of modern and traditional varieties. More systematic efforts and a certain level of organization are necessary to make full use of knowledge already existing within a culture, to exploit fully the potential within the locally available germplasm, and to take advantage of opportunities provided by science.

Community Organization

Widely diverse forms of organizations dealing with seed management have sprouted up at the community level in recent years. I will present an Ethiopian case representing a traditional society, and a Philippino case representing a modern society. Criteria of classifying these societies as traditional and modern are access to external markets and farm inputs. These were absent in the Ethiopian case and present in the Philippino context.

The Ethiopian case is from Tigray, in the middle of the famous Abyssinian gene centre. Renowned for genetic richness and for knowledge and culture related to seed management, the area should be expected to provide an excellent site for community seed banks. But seed banks were organized more because of poor seeds than because of genetic wealth. Poor seeds were seen as one of the reasons for poor agricultural performance. This was in the 1980s and the area was isolated by war. No external support was possible and community leaders had to look for sources of improvement within their own communities. They knew that those sources existed as, in all societies, there were experts known for their skills in traditional seed selection who had fine local seeds. But they needed an organization to extend the benefits of those experts and those seeds to the wider community.

The Philippine case is from Mindanao, from a community where all farmers have formal school education, and where most of them have taken over their farms after the introduction of modern farming. These farmers had no memory of a pre-Green Revolution practices, such as seed selection and traditional seed management. In recent years, however, some of them have switched to organic farming because of declining profit margins in the high-input system. This has created a need for a different type of seed and also an organization to recover and reintroduce the lost traditional seeds as well as to re-establish a local seed system.

These two cases, one isolated from, and the other influenced by, the modern system required different organizational approaches to the seed management problems. In both cases, however, local seeds and on-farm selection were established as the platform on which to build. In Ethiopia, the felt problem was poverty and recurrent famines. Community leaders saw poor crop performance associated with poor seeds as one of the causes, but also knew about individuals who had good seeds and had a reputation of being excellent seed selectors. They decided to extend the good seeds through a credit scheme. Community seed banks were established and the local experts on traditional seed selection were used to identify good seeds for lending. Unlike genebanks, they were not concerned with conservation, but rather with the circulation of seeds. Like a commercial bank, they put their capital to work. They also had social concerns and gave priority to loan applicants who were poor and had a particular need for access to good seeds (and who needed protection against private moneylenders). The seed banks were owned by the community and controlled by democratically-elected community assemblies.

The first of these seed banks was established in 1988 and, within a few years, was replicated in most local districts of a region of close to four million people. The growth of the community seed banks continued after peace, in 1991. But now the challenge is twofold: the seed banks must develop in order to remain relevant; and they must protect their integrity and independence when government institutions and seed companies start appearing in the area.

During the war, people had no choice. They had to depend on their own resources. And they proved to themselves that the necessary skills and the required good seeds existed within their own community and could be used to solve their immediate problems. Currently, these Ethiopians do not seem to consider their seed effort as an emergency measure that can be phased out when government services and seed companies begin to function and, perhaps, to take over. They want to keep their seed banks as permanent institutions. Their challenge now is to realize the development potential of these institutions and the evolutionary potential of their seeds within the context of an opened-up economy. But the seed banks and their associated seed supply system are also a challenge for the scientific plant breeding system which is now being re-established in the area. That challenge need not render the local seed supply system obsolete by supplying better commercial seeds; scientists could opt to work with local farmer-breeders in order to ensure that seeds offered through the seed banks remain competitive.

The case from the Philippines, the Community Based Native Seed Research Centre (CONSERVE) in Mindanao, arose as a response to critical economic problems of the high-input system of rice cultivation. With increasing prices for inputs and decreasing prices for the produce, profit margins were shrinking, and farmers became dangerously dependent on moneylenders. Some individuals saw a switch to organic farming as the only way out. In that situation, farmers needed a new organization. Unlike the Ethiopian case, where community assemblies representing the entire community were the organizers and owners of the seed activities, this Philippine group was a minority, and membership was individual.

Starting in 1992, this group organized a search for local traditional seeds, which were recovered from farms in isolated remote areas. These seeds were multiplied and distributed to the members for on-farm evaluation and screening. After only two years of operation, I visited the project and found farmers discussing seeds with excitement. From an initial challenge of sorting and selecting among a great number of landraces offered to them, some had already started selecting within landraces, and some had started crossing varieties and keeping written records of what they were doing. They involved their wives and children. More than a change of seeds had occurred; there was a change of mind also. Before, these farmers grew modern varieties. Such varieties are supposed to be pure, and off-types are considered as impurities. In case they were saving seeds, they had been taught to rogue the off-types before harvest to maintain varietal integrity. The change of mind involved seeing diversity in the field as a resource, rather than an impurity, and seeing themselves as active selectors, rather than passive receivers of ready-made varieties.

It is a mistake to consider this as a rejection of science. It is a withdrawal from the commercial seed system. If it is true that scientific plant breeders are working for farmers and not for seed companies, they might find organized farmer groups another outlet for their scientific achievements. That would require the development of participatory plant breeding methods.

BREEDING PERENNIAL GRAINS: COOL-SEASON GRASSES

WHEAT

Development of perennials up to 1990: Hexaploid wheat arose approximately 5000 years ago when the genomes of tetraploid wheat (*T. turgidum* or *T.*

carthlicum, both 2n=28, AABB) and an Asian goatgrass (*Aegilops tauschii,* 2n=14, DD) were combined via amphiploidization (*i.e.*, natural hybridization followed by spontaneous production of 2n gametes in the hybid.) Tetraploid wheat itself is a much older natural amphiploid, incorporating the genomes of two diploid grasses. Both wheats are part of the large and diverse Triticeae tribe of the grass family (Gramineae), which also includes scores of perennial species, many of which can be hybridized with wheat. Wagoner examined in detail the early history of efforts in the United States, Canada, Germany, and, most importantly, the USSR, to transfer genes for perenniality from alien grass species into bread wheat, citing more than 65 publications on the subject. None of these efforts produced a truly perennial grain cultivar, but they did spin off much valuable annual germplasm with genes for disease resistances and other traits. In the end, most of the effort in the perennial-wheat programmes was diverted into producing improved annual cultivars, where progress was more easily achieved.

Of the few perennial, grain-producing genotypes developed from wide hybrids at the time of Wagoner's review, none was agronomically successful. Soviet-developed 'perennial' cultivars produced good grain harvests only in the year in which they were established from seed; in the end, they were used mainly as forage cultivars that provided no more than one grain harvest. The US germplasm 'MT-2', derived from a hybrid between *T. turgidum* and *Thinopyrum intermedium* (2n=42), released by Schulz-Schaeffer and Haller in Montana, had very low kernel weight and unreliable persistence. In Sweden, Fatih found that yields of perennial *T. aestivum/Th. intermedium* partial amphiploids (2n=56) were, on average, only 48% of the yields of 42-chromosome, annual, backcross-derived lines of similar parentage.

Production of new hybrids: In the decade since Wagoner's review, no perennial wheat cultivars have been released for production. But basic research on hybridization and cytogenetics has opened up new possibilities for geneticists and breeders interested in the problem.

Species of the genus *Thinopyrum* have been hybridized with wheat more often than have any other perennial species because of the ease of producing partially fertile hybrids, often without embryo rescue. *Thinopyrum intermedium* (2n=42) is rhizomatous, and two other commonly utilized species, *Th. ponticum* (2n=70) and *Th. elongatum* (2n=14) are caespitose. Researchers at Washington State University have launched a new programme to develop perennial wheat from *Thinopyrum* crosses, with promising preliminary results. As we shall, recent cytological and molecular studies have explained why *Thinopyrum* crosses have often led to frustration, and in doing so, have suggested new approaches.

Other species in the genera *Thinopyrum, Elymus,* and *Leymus* have long been investigated as sources of perenniality, and the perennial gene pool available to wheat geneticists is growing rapidly. In a review, Sharma and Gill listed only 16 perennial species that had been hybridized with hexaploid or tetraploid wheat, including only one species — *Elymus giganteus* — outside the *Agropyron-Thinopyrum* complex. A decade later, Jiang *et al.* added 38 additional species, including 17 in *Elymus* and 6 in *Leymus,* to the list of hybrids. Sharma reviewed the production of hybrids between wheat and more than 50 perennial species.

Our ability to make crosses and backcrosses between distantly related species has grown along with advances in embryo rescue, hormone treatments, intra-ovarian fertilization, bridge crosses, and protoplast fusion. The effects of these techniques are magnified when they are used to exploit intraspecific variation for crossability, combining ability, or variation among species carrying the same or similar genomes. Especially important are reciprocal crosses followed by embryo rescue when the traditional method of using the species with higher chromosome number fails.

Within each perennial species, accessions can vary in crossability, and choice of the annual wheat parent can also have a strong effect. For example, homozygosity for kr crossability alleles often makes hybridization possible, and may even improve early seed development; however, it may not affect results in extremely wide crosses. Choice of the wheat parent may depend on other considerations. Hybrids of *Thinopyrum* and *Leymus* with Chinese Spring (*krkr*) could not survive winter temperatures, whereas hybrids with two Japanese spring wheats were winterhardy. Fertility in the F1 through production of unreduced gametes can be induced by crossing the perennial species with the tetraploid wheat *T. carthlicum* and maintaining low temperatures during pollination and embryo development.

Many wheat/*Elymus* hybrids have been made in recent years, and addition, substitution, and translocation lines have been developed. But to our knowledge, no explicit effort to transfer perenniality from *Elymus* to wheat is underway. Lu and von Bothmer produced hybrids between 12 *Elymus* species (2n=28 or 42) and wheat (2n=42). Hybridization was made possible by using *Elymus* as the female parent and rescuing hybrid embryos, as first demonstrated by Sharma and Gill. All hybrids were perennial, but chromosome pairing was very low, as expected. The hybrids were not treated with colchicine to produce amphiploids.

Hybridization between wheat and the genus *Leymus* has a long history, but has not led to perennial cultivars. Hybrids are most easily obtained with the larger-seeded, self-pollinated species of the genus: *L. arenarius* (2n=56), *L. racemosus* (2n=28), and *L. mollis* (2n=28), and these species also have the greatest agricultural potential. Wheat has been crossed with eight *Leymus* species in all. Most crosses require embryo rescue. Hormonal treatment has been used in producing F1 seed. Wheat/*Leymus* hybrids produced by the Soviet perennial-wheat breeding programme were perennial but less winterhardy than wheat-*Thinopyrum* hybrids. In recent studies, hybrids produced by pollinating two wheat species (*T. aestivum* and the tetraploid *T. carthlicum*) with *L. arenarius* and *L. mollis* were perennial, producing short rhizomes and exhibiting some intergenomic pairing and a high enough level of fertility to permit backcrossing. Hybrids between *T. aestivum* and two other species, *L. innovatus* and *L. multicaulis,* were non-rhizomatous.

Advances in chromosome identification: A revolution in cytogenetic techniques has led to a better understanding of the problems faced by perennial wheat breeders. MT-2 provides a good example. It was derived by selfing a 70-chromosome amphiploid containing the genomes of durum wheat (*T. turgidum,* 2n=28, genomes AABB) and *Th. intermedium,* (2n=42, genomes StStEEEStESt). Schulz-Schaeffer and Haller predicted that MT-2 would stabilize at 2n=56 through elimination of *Thinopyrum* chromosomes. In fact, according to Jones *et al.*, individual plants of MT-2 vary in chromosome number, with most having 2n=56.

But genomic *in situ* hybridization (GISH) showed that 56-chromosome MT-2 plants contained numbers of wheat chromosomes varying between 24 and 28. Therefore, there had been loss of wheat chromosomes in some plants, and *Thinopyrum* chromosomes had been eliminated at random. Plants contained up to four St-E or St-Est translocated chromosomes but no wheat-*Thinopyrum* translocations. A wheat/*Thinopyrum* hybrid known as AT 3425, which has resistance to Cephalosporium stripe disease and perennial growth habit, has 2n=56; fluorescent genomic in situ hybridization (FGISH) and C-banding showed that AT 3425 has 36 wheat chromosomes, 14 T*hinopyrum* chromosomes, and 6 chromosomes resulting from translocations between the two species.

The *Thinopyrum* chromatin in AT 3425 and another line with the same chromosome configuration, PI 550713, probably originated from *Th. ponticum,* and both lines are cytologically stable. Another perennial, 56-chromosome line, AgCs, carries the combined genomes of hexaploid wheat and the diploid species *Th. elongatum*.

Banks *et al.* examined meiotic pairing in crosses among eight 56-chromosome partial amphiploids derived from crosses between hexaploid wheat (*T. aestivum*) and *Th. intermedium.* As is often found, the lines had all 42 wheat chromosomes as well as 14 from *Thinopyrum.* The sets of chromosomes originating from *Th. intermedium* differed in all but two lines. Unfortunately, all eight amphiploids studied meiotically by Banks *et al.* were annual; two Soviet-developed perennial wheats were examined only phenotypically. Recently, partial amphiploids that had originated in the Soviet progam have been found to be hexaploid, containing 30 chromosomes from tetraploid wheat and 12 from *L. mollis*. Thus, as in wheat-*Thinopyrum* amphiploids, elimination of the perennial parent's chromosomes had occurred. In addition, one pair of wheat chromosomes had been substituted for a pair from *L. mollis*. The partial amphiploids were annual.

Thinopyrum elongatum (2n=14) is a diploid that is more difficult to hybridize with wheat than are *Th. intermedium* or *Th. ponticum;* however, the resulting amphiploids are vigorous, stable, and perennial. Because *Th. elongatum* contributes only a single genome, all plants should have the same 56-chromosome complement. For example, the wheat/*Th. elongatum* amphiploid AgCs is cytologically stable and perennial. Jauhar produced trigeneric hybrids between a *Th. bessarabicum-Th. elongatum* amphiploid and tetraploid wheat (*T. turgidum*). The hybrids (2n=28, ABJE) were vigorous and perennial.

The maximum number of chromosomes that can be tolerated in amphiploids between wheat and *Thinopyrum* spp. appears to be 56, although 42-chromosome genotypes are more meiotically stable. Usually, partial amphiploids resulting from crosses with tetraploid wheat will contain approximately 28 chromosomes from the wheat parent and approximately 28 from *Thinopyrum,* whereas partial amphiploids with hexaploid wheat will have approximately 42 from wheat and 14 from *Thinopyrum.* Cauderon pointed out that partial amphiploids are automatically selected for "good balance" between wheat and perennial chromosomes through the elimination process; however, as we have seen, the same chromosomal complements are not consistently selected. Breeding populations based on collections of such lines would be plagued with sterility and lack of chromosome pairing unless a diploid perennial parent such as *Th. elongatum* is used.

The goal of a breeding programme cannot be to develop a single perennial wheat cultivar. One partial amphiploid carefully selected to be cultivated as a perennial would have a unique chromosomal constitution. It would be a gene pool of one individual — a dead end in a breeding programme. To launch a perennial wheat breeding programme based on partial amphiploids would be an ambitious undertaking, involving the following steps for any polyploid perennial species targeted:

- Hybridize tetraploid and hexaploid wheats with the polyploid perennial species, making many parental combinations and sampling diversity of all parental species.
- Produce amphiploids and self-pollinate with mild selection for enough generations to achieve stable chromosome numbers.
- Use *in situ* hybridization, chromosome banding, genetic markers, and other techniques to identify chromosomes in a large population of selected plants representing many parental combinations.
- Assign partial amphiploids to groups of homogeneous chromosomal constitution.
- Compare groups for all phenotypic traits of interest and select one or a few on which to base further breeding. Develop foundation

breeding pools from those groups. Experiment with intercrosses between groups, selecting for potentially superior chromosomal combinations.

- In creating new partial amphiploids to introduce into breeding pools, select strictly for appropriate chromosomal complements.

Such a plan would require a mammoth investment of resources, but the initial production of stable partial amphiploids is feasible on a large scale. Selfing and stabilization will occupy several years; by the time genotypes requiring chromosome identification can be produced, vastly more efficient cytological and molecular techniques are almost certain to be available, bringing the breeding programme into the realm of the practical. But technological improvements do not guarantee the development of truly perennial grains, and parallel strategies are needed.

New strategies for perennial wheat: The perennial grasses of the tribe Triticae have long been used in wheat improvement, primarily as sources of individual resistance genes. A traditional strategy for transferring genes is to produce F1 hybrids between wheat and the donor species; either double the chromosome number of the hybrid to produce an amphiploid or pollinate the F1 directly; backcross to wheat genetic stocks to produce lines carrying the normal wheat complement of 42 chromosomes plus one or a pair of chromosomes from the donor parent; and — at some point in the process — attempt to induce a translocation that transfers a segment carrying the target gene to a wheat chromosome. There are many variations on this strategy, but the usual goal is to transfer a single gene, eliminating as much of the rest of the donor genome as possible. Perenniality in wheat's relatives is more genetically complex than the single-gene traits transferred to date and may require a different approach. Amphiploids are not always perennial, and as we have seen, they are usually genetically unstable and agronomically undesirable. Backcrossing to wheat usually results in a return to the annual habit. Therefore, Anamthawat-Jonsson proposed backcrossing instead to the perennial parent — in their research, either *Leymus arenarius* or *L. mollis.* The objective then becomes to improve traits such as grain yield and kernel weight in the perennial species.

Anamthawat-Jonsson listed the traits to be improved by incorporating wheat germplasm into *Leymus* species: perenniality, grain quality, harvestability, threshability, kernel weight, synchronization of maturity, lodging and shattering resistance, meiotic stability, and, of course, grain yield. She has backcrossed partial amphiploids to *L. mollis,* and the progeny were vigorous with long rhizomes. In light of the many problems that have been encountered in transferring perenniality to wheat, the converse approach — using wheat to improve the perennial species — may have considerable merit. Backcrossing to the perennial will almost certainly produce breeding populations with low average grain yields and a high frequency of shattering, requiring the screening of large numbers of genotypes. Logic suggests that perennial allopolyploid species will be more tolerant of added or substituted wheat chromosomes. *L. arenarius* has been used as a grain crop in the past and efforts to improve it using wheat as a donor parent are underway. *Th. ponticum* is easily crossed with wheat but its high ploidy level and lack of diploidization probably would prevent its use in grain production. Wheat might be used as a donor parent for improving *Th. intermedium* or some hexaploid species of *Elymus* with which it can be crossed.

B. Rye

Diploids: Rye (*Secale cereale*) appears to be at least as promising a candidate for perennialization as is wheat. Its chromosomes are homologous with those of its direct perennial ancestor, *S. montanum.* Both species are diploid and cross-pollinated. Rye is very winterhardy, well adapted for grazing, and useful in weed control because of its allelopathic properties. The Soviet perennial-grains programme included a large effort in rye, and they produced some weakly perennial genotypes that were used in limited production. Later, a decades-long effort to breed perennial rye in Germany met with only partial success. Recently, a perennial rye cultivar, 'Perenne', was released in Hungary for grain and forage production.

Despite initial expectations, no perennial rye cultivar has been used in full-scale grain production. Breeders have been stymied by the tendency of plants in *S. cereale*/*S. montanum* populations to be either fertile and annual or highly sterile and perennial. A chain of translocations involving three of rye's seven pairs of chromosomes separates the two species, and gene(s) from *S. montanum* governing perenniality are located on one or more of the translocated chromosomes. Because meiosis in plants heterozygous for one or more translocations produces many inviable gametes with duplications or deficiencies of chromosomal segments, plants in interspecific rye populations fall into one of three categories: homozygous for the *S. cereale* chromosomal arrangement (fertile, annual); heterozygous for one or more of the translocations (highly sterile); or homozygous for the *S. montanum* arrangement (perennial, fertile).

Plants in this last category would seem to answer the breeder's need; however, they are rare, and the large portion of their genomic content derived from *S. montanum* reduces their agronomic desirability and spike fertility. According to Reimann-Philipp, seed-set in *S. montanum* itself is low — approximately 80%. Reimann-Philipp selected intensely for the *S. cereale* phenotype within an interspecific, perennial population. This population was presumed to be homozygous for the three *S. montanum*-derived chromosomes involved in the translocations, identified as $4R^{mon}$, $6R^{mon}$, and $7R^{mon}$ by Koller and Zeller but referred to as $2R^{mon}$, $6R^{mon}$, and $7R^{mon}$ by Reimann-Philipp. He was attempting to keep these chromosomes fixed while restoring completely the other four chromosome pairs from *S. cereale* through recombination and selection. But he could not achieve a kernel weight greater than 15 mg.

As an alternative, Reimann-Philipp proposed selection for perennial plants carrying chromosomes 2R, 6R, and 7R of *S. cereale.* Presumably, such plants would arise from recombination within the ring of six translocated chromosomes. Dierks and Reimann-Philipp had postulated that perenniality was governed by a single gene that lay approximately 10 crossover units from one of the breakpoints. Selection for perenniality would be routine, but selection for the *S. cereale* chromosomal constitution would require either a laborious testcross procedure or a cytological test. A morphological difference between chromosomes 6R and $6R^{mon}$ did not prove satisfactory for this purpose. Today, the extensive genetic map of rye could allow marker-assisted selection for the *S. cereale* arrangement. A large initial experiment could provide much more detailed information on the genetic control of perenniality; as noted by Reimann-Philipp, the trait is probably affected by more than one gene.

Yet another strategy was followed by L.F. Myers and R.J. Kirchner in the breeding of 'Black Mountain' perennial rye in Australia: backcrossing the interspecific hybrid twice to the *S. montanum* parent. Perennialism (and, presumably, the *S. montanum*-type chromosomal arrangement) was quickly restored by backcrossing. But this cultivar was intended primarily as a forage grass, with *S. cereale* donating genes for nonshattering rachis and improved seed production. Oram practiced six cycles of half-sib family selection for grain and forage yield in Black Mountain, achieving gains in both traits while maintaining a low level of shattering. With grazing, stands of 'Black Mountain' decline after 3 years; however, if shattering is permitted, stands can be continually replenished by volunteer seedlings. Without selection to develop a cultivar strictly for grain production, we cannot know whether backcrossing to perennial rye while selecting for alleles from annual rye can achieve sufficient yield improvement.

Reimann-Philipp warned of a hazard when growing diploid perennial rye with the *S. montanum* chromosomal arrangement on a field scale. If pollen from the perennial drifted into seed production fields or breeding nurseries of annual rye, the resulting translocation heterozygosity would seriously and irreversibly degrade fertility in subsequent generations.

Tetraploids: In an effort to improve kernel weight, Reimann-Philipp used colchicine to double the chromosome number of a perennial *S. cereale/S. montanum* population homozygous for the $4R^{mon}$, $6R^{mon}$, and $7R^{mon}$ chromosomes. The resulting tetraploid, named 'Permontra', had a kernel weight of approximately 30 mg (double that of the diploid), and first-year grain yields over 2000 kg/ha when grown in Germany. Yields declined in subsequent years. 'Permontra' achieved a similar grain yield in the Land Institute's plots in Kansas, but only 15 to 20% of the plants regrew in the next season. After a first-year harvest in the hot, dry summer of 2001, a stand of Permontra at The Land Institute died out completely by September.

Another problem with 'Permontra' — poor seed set — is also common in annual tetraploid rye, because the formation of multivalent chromosomal associations leads to production of gametes with extra or missing chromosomes. Selection can improve seed-set in tetraploid rye, and Reimann-Philipp pointed out that selection for seed set or meiotic stability can be practiced much more effectively in a perennial, by screening phenotypically or cytologically in one flowering cycle and intercrossing selected plants in the next. He found that phenotypic selection greatly improved seed-set in the perennial spring rye 'Soperta', which was derived from seven 'Permontra' plants that did not require vernalization in order to flower. Another approach would be to introduce the *Ph1* gene from wheat into a tetraploid hybrid between *S. cereale* and *S. montanum* to enforce diploid pairing and improve fertility. The *Ph1* gene was shown to operate when the chromosome carrying it was added to rye.

Tetraploid perennials may provide advantages beyond increased kernel weight. They are reproductively isolated from diploid annual seed production fields. In one study, 'Permontra' had greater heat and drought tolerance, and a much more extensive root system, than did diploid or tetraploid annuals. But perennial rye, whether tetraploid or diploid, will not be grown widely as a grain crop until the problems of sterility, persistence, and maintenance of yield over seasons are solved.

Triticale: To date, the only species to be synthesized by artificial hybridization for use as a cereal crop is triticale (X *Triticosecale*), an amphiploid of durum wheat (*T. turgidum*) and *S. cereale.* (Octoploid triticale cultivars — *T. aestivum/S. cereale*— have also been produced but not widely used commercially.) Triticale has not become one of the world's leading cereals, but its modest success suggests the possibility of developing perennial *T. turgidum/ S. montanum* triticales. Derzhavin produced and intercrossed many amphiploids derived from crosses between durum wheat and perennial rye accessions. He augmented the gene pool by allowing the amphiploids to pollinate a large number of different wheat/rye F1 hybrids. But the resulting populations were only weakly perennial. Three-way hybrids — from crosses between wheat/*Th. intermedium* hybrids and perennial rye— were more strongly perennial but had sterility, low yields, and small seeds.

Robert Metzger (USDA-ARS retired, Corvallis, OR, personal communication) reports that a *S. montanum*-derived triticale that he has developed is not sufficiently perennial, but he recommends producing and screening more new amphiploids involving a wider range of *S. montanum* germplasm. It could be that no triticale will be fully perennial, having only one of three genomes derived from a perennial species. Intercrossing of diverse lines followed by selection could improve persistence over years. If improved hexaploid or tetraploid perennial wheats can be developed, they could be crossed with *S. montanum* to produce more strongly perennial triticales.

In attempting to develop perennial triticales, breeders can take lessons from development of the annual crop. Primary triticales, *i.e.*, newly doubled wheat/rye hybrids, inevitably suffer from sterility, seed shrivelling, lodging, and low yield potential.

Decades of intense selection and introgression have resulted in triticales that are cytologically stable and improved for all of these traits, but they stand on a narrow germplasm base. Improving the performance and genetic variability of the triticale gene pool can be accomplished by several means: production of new primary triticales; triticale/wheat crosses; triticale/rye crosses; and crosses between hexaploid and tetraploid triticales.

Primary triticales derived from the wild *S. montanum* are even more agronomically primitive than *S. cereale*-based primary triticales and will require even greater breeding effort with a wide range of parents. "Substituted" triticales in which one or more wheat chromosomes replace those of rye are often agronomically superior; chromosomes 2D and 6D appear to be selectively propagated by breeders in populations segregating for R-and D-genome chromosomes.

But it must be kept in mind that until genes conditioning perenniality can be mapped, random substitution of wheat for *S. montanum* chromosomes will reduce the chances of selecting a strongly perennial triticale.

Direct Domestication of the Perennial Triticeae

Intermediate wheatgrass: From domestication of perennial grasses with wheat or rye as a donor parent, it is a relatively short leap to domestication without any interspecific crossing. Three large-seeded perennial species that have been hybridized with wheat also have attracted attention as candidates for direct domestication. By far, the most work in this area has been done with *Th. intermedium* by Wagoner at the Rodale Institute in Pennsylvania and her colleagues at the USDA-NRCS Big Flats Plant Materials Centre in New York. Wagoner described in detail the characteristics that make intermediate wheatgrass a good candidate for domestication as a perennial grain, while noting shortcomings that must be addressed. Becker *et al.* concluded that its grain has protein quality "superior to the cereal grains now commonly grown", with no significant amounts of antinutrients. Recurrent selection is a logical breeding method for improving an only slightly domesticated, cross-pollinated species like *Th. intermedium.* Using mass selection without controlled pollination, Knowles increased seed yield in an intermediate wheatgrass population by 10% per cycle. In each cycle, 1000 plants were evaluated for spike fertility over three years, and the best 50 were selected. When selected plants were removed to the greenhouse over the winter to exclude pollination by non-selected plants, thereby doubling parental control, gain per cycle increased to 20% — a result perfectly consistent with selection theory. Gridded mass selection is used to exercise control over microenvironmental effects and increase selection response. Wagoner evaluated 300 accessions of intermediate wheatgrass, for grain yield, yield components, and end-use quality, selecting the 20 best accessions in 1989. The selections were transplanted into a polycross nursery, and 380 progeny resulting from pollination among the selections were evaluated, in a field divided into blocks of 25 plants each, between 1991 and 1994.

The best 11 plants resulting from within-and among-block selection, plus three selections resulting from further evaluation of other accessions, were put into a second-cycle polycross, and 400 individual progenies were evaluated in a second blocked nursery. Yield per plant in the 14 selections was approximately 25% higher than the population mean. Evaluation of the second-cycle population is underway, and selected plants will be intermated in 2002 to complete another breeding cycle (M. van der Grinten, USDA-NRCS, Big Flats, NY, pers. commun.) Better environmental control through selection within blocks may have produced the five-percentage-point improvement in selection response over that of Knowles but it remains to be seen if the small effective population size in the second cycle (14 plants) will restrict genetic gain in the future.

Wildrye: The Land Institute in Kansas has studied perennial cool-season grasses as potential grain crops for over 20 years. They evaluated almost 1500 accessions representing 85 species of *Agropyron, Thinopyrum, Elymus,* and *Leymus,* along with 2630 accessions of other species, between 1979 and 1987. The species selected as having the greatest potential for domestication was *L. racemosus,* known commonly as giant or mammoth wildrye. However, prospects for utilization of this species in the near future are unclear. Among 16 accessions evaluated over 2 years, yields did not exceed 830 kg/ha, and yield declined rapidly in the second and third years. Wildrye's great vigour, accompanied by large spikes but sparse seed-set, resulting in low harvest index, may provide considerable scope for breeders to select for diversion of photosynthate towards grain production. But there is no current breeding programme for grain yield in *L. racemosus;* until selection is undertaken, no conclusions can be drawn regarding its potential. *L. racemosus* is self-pollinated and would require a breeding approach different from that taken with *Th. intermedium.*

Lyme grass: Lyme grass or beach wildrye (*Leymus arenarius*) has been used as a food grain since the time of the Vikings, and, as we have seen, is being studied as a potential grain crop in Iceland. There is significant genetic variation among accessions of *L. arenarius* and *L. mollis*, and it would be interesting to know which approach would result in more rapid genetic progress: direct selection within the species or an interspecific backcross programme using wheat as a donor parent. The latter strategy takes advantage of genes selected through millenia of wheat domestication and breeding, but introduces chromosomal instability.

Prospects for direct domestication: Is there sufficient genetic variation within these three, or other, cool-season grasses to support large improvements in yield, kernel weight, and other traits? The very existence of annual grain crops proves that selection over thousands of years can move the mean of a species far beyond its original phenotypic range. Gains of 20 to 25% per cycle are much more rapid than typical gains in major annual crops, even considering the longer selection cycle of perennials. But to effect sufficient changes in a matter of decades rather than centuries — while possibly working "uphill" against the problem of resource allocation in perennials — will require much larger breeding efforts than have been undertaken to date. Relatively small efforts at domestication, which are within the capabilities of nonprofit organizations such as the Rodale Institute or Land Institute (or small-scale breeding programmes within larger organizations such as USDA or universities), must be expanded to a much larger scale by university, government, or corporate breeding programmes if wholly new perennial grain crops are to be developed.

The yield increases of 25% per cycle achieved by Wagoner *et al.* are remarkable, especially considering that they selected for other traits in addition to yield. But because response to recurrent selection tends either to follow a linear path or decelerate, future gains per cycle will probably be no greater than a constant percentage of the base population's yield. If a hypothetical perennial grass population yielding 500 kg/ha of grain undergoes selection, with a yield increase of 125 kg/ha/cycle (25% of the base yield), 20 cycles will be required to reach 2500 kg/ha. Because selection in perennials must be based on evaluation over two or more seasons, a single cycle can occupy four or five years. Obviously, if it is going to take almost a century to develop a high-yielding perennial crop through direct selection, a long-term commitment is required; however, such a rate of progress is much greater than the rate at which our annual crops were domesticated and improved. Marker-assisted selection and/or some genetic input from wheat could speed up the process. *Oat*: Perennial oats for grain production might be developed from crosses between the cultivated hexaploid oat (*Avena sativa,* 2n=42, genomes AACCDD) and a wild, perennial, autotetraploid relative, *A. macrostachya* (2n=28, CCCC). Such crosses require embryo rescue.

The F1 is highly sterile, but backcrosses to *A. sativa* have been made, and limited pairing between chromosomes from different parents does occur. J. P. Murphy has produced a 70-chromosome amphiploid between the species. The objective of this cross is to improve winterhardiness in the annual crop; perenniality has not been evaluated. Because of the partial homology that exists between chromosomes of the parental species, the amphiploid is likely to suffer from the same chromosomal instability found in wheat amphiploids. But the amphiploid, like the hybrid, can be backcrossed to *A. sativa* (Murphy, pers, commun.) Ladizinsky domesticated accessions of two wild annual oat species by using the cultivated oat A. sativa as the donor of genes for nonshattering and other traits. Perhaps this approach could be tried with *A. macrostachya.*

More hybrid combinations and larger populations will be needed if genetic studies and selection for perenniality are to succeed in the backcross generations. A more diverse sample of *A. macrostachya* parents would be desirable, but the species is restricted to two mountain ranges in Algeria, limited germplasm collections exist in the United States, and there are very few accessions held in other countries.

Selection for perenniality in colder climates could be thwarted by the lack of winterhardiness in oats; winter annual oats are not generally sown above 35 degrees latitude in North America. At sites where *A. macrostachya* was collected by Guarino *et al.*, the mean minimum temperature of the coldest month ranged from-0.6 to-3.6°C. *Rice*: Although tropically adapted, rice (*Oryza sativa*) has the C3 carbon fixation pathway and is included here with the cool-season grasses. The perennial ancestor of *O. sativa* is *O. rufipogon.* Both species are diploid (2n=24) with homologous chromosomes and they can be hybridized easily. Indeed, natural hybridization and introgression occur in the field. From 1995 to 2001, the International Rice Research Institute (IRRI) had a programme for development of perennial rice cultivars to reduce erosion on the steep slopes where upland rice is often grown. Populations from IRRI's breeding programme, which was discontinued in 2001, have been distributed to cooperators in China, where perennial rice breeding efforts continue.

Sacks *et al.* found wide variation in second-year survival among 51 *O. sativa/ O. rufipogon* F1 hybrids. Sixteen percent of the hybrid combinations had greater than 50% survival, and 19% of all hybrid plants survived. In a cross between a rice cultivar with a regeneration score of 1.0 and an accession of *O. rufipogon* with a score of 3.8, the F1 had a regeneration score of 4.0, and the scores of F2 clones ranged from 0 to 5. Paradoxically, in three of the four chromosomal segments that affected regeneration ability, it was the annual parent's allele that had a positive effect. In contrast to mapped regrowth loci in *Sorghum bicolor/ S. propinquum* populations, none of the regeneration loci were associated with effects on tiller number.

Selection in interspecific populations may be aided by rice's detailed molecular map and the known locations of chromosomal segments affecting traits of domestication. Many of the traits separating the annual and perennial species show polygenic inheritance. Surprisingly, *O. rufipogon* was the source of four chromosomal segments with positive effects on testcross grain yield in one set of backcrosses; however, three of the four segments were adjacent to segments that either increased plant height or delayed maturity. With positive alleles affecting perenniality *and* productivity apparently being contributed by *both* parental species, prospects for breeding high-yielding perennial rice genotypes may be bright.

Oryza sativa can also be hybridized with *O. longistaminata,* the perennial ancestor of West African rice, *O. glaberrima.* The perenniality of *O. rufipogon* lies in its ability to regrow repeatedly through production of new tillers, whereas *O. longistaminata* regrows from rhizomes. In crosses between *O. sativa* and *O. longistaminata,* genes affecting rhizome production appear to be linked to genes for hybrid embryo abortion. Consequently, IRRI scientists backcrossed rare hybrids to both parental species and intercrossing the progeny in an effort to develop a rhizomatous, agronomically acceptable genotype. Tao *et al.* recovered a single rhizomatous individual from among 162 plants produced by backcrossing an *O. sativa/O. longistaminata* hybrid to *O. sativa.*

A long-lived, three-species hybrid (*O. sativa/O. rufipogon/O. longistaminata*) has persisted through winters in China with monthly mean temperatures as low as 5°C. Crossing the hybrid with *O. sativa* and intermating perennial progenies has eliminated shattering. One possibility for breeding an even more cold-tolerant perennial rice exists. An ecotype of *O. rufipogon* known as 'Dongxiang' has the ability to regrow in regions of China where temperatures below -10°C are common.

2

Potato

The potato is a starchy, tuberous crop from the perennial nightshade *Solanum tuberosum* L. The word "potato" may refer either to the plant itself or to the edible tuber. In the Andes, where the species is indigenous, there are some other closely related cultivated potato species. Potatoes were introduced outside the Andes region approximately four centuries ago, and have since become an integral part of much of the world's food supply. It is the world's fourth-largest food crop, following maize, wheat, and rice.

Wild potato species occur throughout the Americas from the United States to southern Chile. The potato was originally believed to have been domesticated independently in multiple locations, but later genetic testing of the wide variety of cultivars and wild species proved a single origin for potatoes in the area of present-day southern Peru and extreme northwestern Bolivia (from a species in the *Solanum brevicaule* complex), where they were domesticated approximately 7,000–10,000 years ago. Following centuries of selective breeding, there are now over a thousand different types of potatoes. Over 99% of the presently cultivated potatoes worldwide descended from varieties that originated in the lowlands of south-central Chile, which have displaced formerly popular varieties from the Andes.

However, the local importance of the potato is variable and changing rapidly. It remains an essential crop in Europe (especially eastern and central Europe), where per capita production is still the highest in the world, but the most rapid expansion over the past few decades has occurred in southern and eastern Asia. As of 2007 China led the world in potato production, and nearly a third of the world's potatoes were harvested in China and India.

ETYMOLOGY

The English word *potato* comes from Spanish *patata* (the name used in Spain). The Spanish Royal Academy says the Spanish word is a compound of the Taino *batata* (sweet potato) and the Quechua *papa* (potato). The name potato originally referred to a type of sweet potato although the two plants are not closely related; in many of the chronicles detailing agriculture and plants, no distinction is made between the two. The 16th-century English herbalist John Gerard used the terms "bastard potatoes" and "Virginia potatoes" for this species, and referred to sweet potatoes as "common potatoes". Potatoes are occasionally referred to as "Irish potatoes" or "white potatoes" in the United States, to distinguish them from sweet potatoes.

The name spud for a small potato comes from the digging of soil (or a hole) prior to the planting of potatoes. The word has an unknown origin and was originally (c. 1440) used as a term for a short knife or dagger, probably related to Dutch *spyd* or the Latin "spad-" a word root meaning "sword"; cf. Spanish "espada", English "spade" and "spadroon". The word spud traces back to the 16th century. It subsequently transferred over to a variety of digging tools. Around 1845, the name transferred to the tuber itself. The origin of the word "spud" has erroneously been attributed to a 19th-century activist group dedicated to keeping the potato out of Britain, calling itself The Society for the Prevention of an Unwholesome Diet. It was Mario Pei's 1949 *The Story of Language* that can be blamed for the word's false origin. Pei writes, "the potato, for its part, was in disrepute some centuries ago. Some Englishmen who did not fancy potatoes formed a Society for the Prevention of Unwholesome Diet. The initials of the main words in this title gave rise to spud." Like most other pre-20th century acronymic origins, this is false.

CHARACTERISTICS

Potato plants are herbaceous perennials that grow about 60 cm (24 in) high, depending on variety, with the leaves dying back after flowering, fruiting and tuber formation. They bear white, pink, red, blue, or purple flowers with yellow stamens. In general, the tubers of varieties with white flowers have white skins, while those of varieties with colored flowers tend to have pinkish skins. Potatoes are mostly cross-pollinated by insects such as bumblebees, which carry pollen from other potato plants, though a substantial amount of self-fertilizing occurs as well. Tubers form in response to decreasing day length, although this tendency has been minimized in commercial varieties.

Fig. Potato plants

After flowering, potato plants produce small green fruits that resemble green cherry tomatoes, each containing about 300 seeds. Like all parts of the plant except the tubers, the fruit contain the toxic alkaloid solanine and are therefore unsuitable for consumption. All new potato varieties are grown from seeds, also called "true potato seed", "TPS" or "botanical seed" to distinguish it from seed tubers. New varieties grown from seed can be propagated vegetatively by planting tubers, pieces of tubers cut to include at least one or two eyes, or cuttings, a practice used in greenhouses for the production of healthy seed tubers. Plants propagated from tubers are clones of the parent, whereas those propagated from seed produce a range of different varieties.

GENETICS

There are about 5,000 potato varieties worldwide. Three thousand of them are found in the Andes alone, mainly in Peru, Bolivia, Ecuador, Chile, and Colombia. They belong to eight or nine species, depending on the taxonomic school. Apart from the 5,000 cultivated varieties, there are about 200 wild species and subspecies, many of which can be cross-bred with cultivated varieties. Cross-breeding has been done repeatedly to transfer resistances to certain pests and diseases from the gene pool of wild species to the gene pool of cultivated potato species. Genetically modified varieties have met public resistance in the United States and in the European Union.

The major species grown worldwide is *Solanum tuberosum* (a tetraploid with 48 chromosomes), and modern varieties of this species are the most widely cultivated. There are also four diploid species (with 24 chromosomes): *S. stenotomum*, *S. phureja*, *S. goniocalyx*, and *S. ajanhuiri*. There are two triploid species (with 36 chromosomes): *S. chaucha* and *S. juzepczukii*. There is one pentaploid cultivated species (with 60 chromosomes): *S. curtilobum*.

There are two major subspecies of *Solanum tuberosum*: *andigena*, or Andean; and *tuberosum*, or Chilean. The Andean potato is adapted to the short-day conditions prevalent in the mountainous equatorial and tropical regions where it originated; the Chilean potato, however, native to the Chiloé Archipelago, is adapted to the long-day conditions prevalent in the higher latitude region of southern Chile.

The International Potato Center, based in Lima, Peru, holds an ISO-accredited collection of potato germplasm. The international Potato Genome Sequencing Consortium announced in 2009 that they had achieved a draft sequence of the potato genome. The potato genome contains 12 chromosomes and 860 million base pairs, making it a medium-sized plant genome.

More than 99 percent of all current varieties of potatoes currently grown are direct descendants of a subspecies that once grew in the lowlands of south-central Chile. Nonetheless, genetic testing of the wide variety of cultivars and wild species affirms that all potato subspecies derive from a single origin in the area of present-day southern Peru and extreme northwestern Bolivia (from a species in the *Solanum brevicaule* complex).

Most modern potatoes grown in North America arrived through European settlement and not independently from the South American sources; however, at least one wild potato species, *Solanum fendleri*, is found as far north as Texas and is used in breeding for resistance to a nematode species that attacks cultivated potatoes. A secondary center of genetic variability of the potato is Mexico, where important wild species that have been used extensively in modern breeding are found, such as the hexaploid *Solanum demissum*, as a source of resistance to the devastating late blight disease. Another relative native to this region, *Solanum bulbocastanum*, has been used to genetically engineer the potato to resist potato blight.

Potatoes yield abundantly with little effort, and adapt readily to diverse climates as long as the climate is cool and moist enough for the plants to gather sufficient water from the soil to form the starchy tubers. Potatoes do not keep very well in storage and are vulnerable to molds that feed on the stored tubers and quickly turn them rotten, however: crops such as grain can be stored for several years with a low risk of rot. The yield of Calories per acre (about 9.2 million) is higher than that of maize (7.5 million), rice (7.4 million), wheat (3 million), or soybean (2.8 million).

HISTORY

The potato was first domesticated in the region of modern-day southern Peru and extreme northwestern Bolivia between 8000 and 5000 BCE. It has since spread around the world and become a staple crop in many countries.

SOUTH AMERICA

Cultivation of potatoes in South America may go back 10,000 years. Yet the tubers do not preserve well in archaeological record, and there are problems with exact identification of those that are found. The earliest archaeologically verified potato tuber remains have been found at the coastal site of Ancon (central Peru), dating to 2500 BC. Also, at Huaynuma, in the Casma Valley of Peru, potatoes dating to about 2000 BC have been found. Early potatoes dating to 800-500 BC were also uncovered at the Altiplano site of Chiripa on the east side of Lake Titicaca.

In the Altiplano, potatoes provided the principal energy source for the Inca Empire, its predecessors, and its Spanish successor. In Bolivia and Peru above 10,000 feet altitude, tubers exposed to the cold night air turned into chuño; when kept in permanently frozen underground storehouses, chuño can be stored for years with no loss of nutritional value. The Spanish fed chuño to the silver miners who produced vast wealth in the 16th century for the Spanish government. Potato was the staple food of most Pre–Columbian Mapuches, "specially in the southern and coastal [Mapuche] territories where maize did not reach maturity".

EUROPE

Sailors returning from the Andes to Spain with silver presumably brought maize and potatoes for their own food on the trip. Historians speculate that leftover tubers (and maize) were carried ashore and planted: "We think that the potato arrived some years before the end of the 16th century, by two different ports of entry: the first, logically, in Spain around 1570, and the second via the British isles between 1588 and 1593... we find traces of the transport of potatoes travelling from the Canaries to Antwerp in 1567... we can say that the potato was introduced there [the Canary islands] from South America around 1562... the first written mention of the potato [is]... a receipt for delivery dated 28 November 1567 between Las Palmas in the Grand Canaries and Antwerp." In 1553, in the book *Crónica del Peru*, Pedro Cieza de León mentions he saw it in Quito, Popayán, and Pasto in 1538.

Basque fishermen from Spain used potatoes as ships' stores for their voyages across the Atlantic in the 16th century, and introduced the tuber to western Ireland, where they landed to dry their cod. The English privateer Sir Francis Drake, returning from his circumnavigation, or Sir Walter Raleigh's employee Thomas Harriot are commonly credited with introducing potatoes into England. In 1588, botanist Carolus Clusius made a painting of what he called "Papas Peruanorum" from a specimen in the Low Countries; in 1601 he reported that potatoes were in common use in northern Italy for animal fodder and for human consumption.

The Spanish had an empire across Europe, and brought potatoes for their armies. Peasants along the way adopted the crop, which was less often pillaged by marauding armies than above-ground stores of grain. Across most of northern Europe, where open fields prevailed, potatoes were strictly confined to small garden plots because field agriculture was strictly governed by custom that prescribed seasonal rhythms for plowing, sowing, harvesting and grazing animals on fallow and stubble.

This meant that potatoes were barred from large-scale cultivation because the rules allowed only grain to be planted in the open fields. In France and Germany government officials and noble landowners promoted the rapid conversion of fallow land into potato fields after 1750. The potato thus became an important staple crop in northern Europe. Famines in the early 1770s contributed to its acceptance, as did government policies in several European countries and climate change during the Little Ice Age, when traditional crops in this region did not produce as reliably as before. At times and places when and where most other crops failed, potatoes could still typically be relied upon to contribute adequately to food supplies during colder years.

In France, at the end of the 16th century, the potato had been introduced to the Franche-Comté, the Vosges of Lorraine and Alsace. By the end of the 18th century it was written in the 1785 edition of *Bon Jardinier*: "There is no vegetable about which so much has been written and so much enthusiasm has been shown... The poor should be quite content with this foodstuff." It had widely replaced the turnip and rutabaga by the 19th century.

French physician Antoine Parmentier studied the potato intensely and in *Examen chymique des pommes de terres* (Paris, 1774) showed their enormous nutritional value. King Louis XVI and his court eagerly promoted the new crop, with Queen Marie Antoinette even wearing a headdress of potato flowers at a fancy dress ball. The annual potato crop of France soared to 21 million hectoliters in 1815 and 117 million in 1840, allowing a concomitant growth in population while avoiding the Malthusian trap. Although potatoes had become widely familiar in Russia by 1800, they were confined to garden plots until the grain failure in 1838–1839 persuaded peasants and landlords in central and northern Russia to devote their fallow fields to raising potatoes.

Potatoes yielded from two to four times more calories per acre than grain did, and eventually came to dominate the food supply in eastern Europe. Boiled or baked potatoes were cheaper than rye bread, just as nutritious, and did not require a gristmill for grinding. On the other hand cash-oriented landlords realized that grain was much easier to ship, store and sell, so both grain and potatoes coexisted.

Throughout Europe, the most important new food in the 19th century was the potato, which had three major advantages over other foods for the consumer: its lower rate of spoilage, its bulk (which easily satisfied hunger), and its cheapness. The crop slowly spread across Europe, such that, for example, by 1845 it occupied one-third of Irish arable land. Potatoes comprised about 10% of the caloric intake of Europeans. Along with several other foods that either originated in the Americas or were successfully grown or harvested there, potatoes sustained European populations.

In Britain, the potato promoted economic development by underpinning the Industrial Revolution in the 19th century. It served as a cheap source of calories and nutrients that was easy for urban workers to cultivate on small backyard plots. Potatoes became popular in the north of England, where coal was readily available, so a potato-driven population boom provided ample workers for the new factories. Marxist Friedrich Engels even declared that the potato was the equal of iron for its "historically revolutionary role". The Dutch potato-starch industry grew rapidly in the 19th century, especially under the leadership of entrepreneur Willem Albert Scholten (1819–92).

In Ireland, the expansion of potato cultivation was due entirely to the landless laborers, renting tiny plots from landowners who were interested only in raising cattle or in producing grain for market. A single acre of potatoes and the milk of a single cow was enough to feed a whole Irish family a monotonous but nutritionally adequate diet for a healthy, vigorous (and desperately poor) rural population. Often even poor families grew enough extra potatoes to feed a pig that they could sell for cash.

A lack of genetic diversity from the low number of varieties left the crop vulnerable to disease. In the early 1800s, a strain of potato blight (*Phytophthora infestans*) known as HERB-1 began to spread in the Americas, especially Central and North America destroying many crops. The blight spread to Europe in the 1840s where, because of an extreme lack of genetic diversity, the potato crops were even more susceptible. In Northern Europe there were major crop losses lasting throughout the rest of the 19th century. Ireland in particular, because of the extreme dependence of the poor, especially western Ireland, on this single staple crop, was devastated by the blight's arrival in 1845.

The Lumper potato, widely cultivated in western and southern Ireland before and during the great famine, was bland, wet, and poorly resistant to the potato blight, but yielded large crops and usually provided adequate calories for peasants and laborers. Heavy dependence on this potato led to disaster when the blight quickly turned harvest-ready and newly harvested potatoes into a putrid mush.

The Irish Famine in the western and southern parts of Ireland between 1845–49 was a catastrophic failure in the food supply that led to approximately a million deaths from famine and (especially) diseases that attacked weakened bodies, and to massive emigration to Britain, the U.S., Canada and elsewhere. During the famine years roughly one million Irish emigrated, this tide was not turned until the 20th century, when Ireland's population stood at less than half of the pre-famine level of 8 million.

AFRICA

It is generally believed that potatoes entered Africa with colonists, who consumed them as a vegetable rather than as a staple starch. Shipping records from 1567 show that the first place outside of Central and South America where potatoes were grown were the Canary Islands. As in other continents, despite its advantages as an anti-famine, high-elevation alternative to grain, potatoes were first resisted by local farmers who believed they were poisonous. Colonialists also promoted them as a low cost food and so it was a symbol of domination. In former European colonies of Africa, potatoes were initially consumed only occasionally, but increased production made them a staple in certain areas. Potatoes tended to become more popular in wartime due to their being able to be stored in the ground. In present day Africa they have become a vegetable or co-staple crop.

In higher regions of Rwanda, potatoes have become a new staple food crop. Prior to the 1994 Rwandan genocide consumption was as high as 153 to 200 kg per year – higher than in any Western European country. Recently farmers have developed the potato as a cash crop after introducing several new varieties brought back by migrant laborers from Uganda and other varieties from Kenya.

ASIA

The potato diffused widely after 1600, becoming a major food resource in Europe and East Asia. Following its introduction into China toward the end of the Ming dynasty, the potato immediately became a delicacy of the imperial family. After the middle period of the Qianlong era (1735–96) in the Qing dynasty, population increases and a subsequent need to increase grain yields coupled with greater peasant geographic mobility led to the rapid spread of potato cultivation throughout China, and it was acclimated to local natural conditions.

Boomgaard (2003) looks at the adoption of various root and tuber crops in Indonesia throughout the colonial period and examines the chronology and reasons for progressive adoption of foreign crops: sweet potato, Irish potato, bengkuang (yam beans), and cassava. Edward Terry mentioned the potato in his travel accounts of the banquet at Ajmer by Asaph Khan to Sir Thomas Roe, the British Ambassador in 1675. It is the earliest mention in history of India. The vegetables gardens of Surat and Karnataka had potatoes as mentioned in Fyer's travel record of 1675. The Portuguese introduced potatoes, the called it 'Batata', in India in early seventeenth century when they cultivated it along the western coast of India. The British traders introduced potatoes in Bengal as a root crop, 'Alu'. By the end of eighteenth century, it was cultivated across northern hill areas of India. Potatoes were introduced to Tibet by nineteenth century through trade route from India.

The potato was introduced in the Philippines during the late 16th century, and to Java and China during the 17th century. It was well established as a crop in Africa by the mid-20th century.

NORTH AMERICA

Potatoes were planted in Idaho as early as 1838; by 1900 the state's production exceeded a million bushels (about 27,000 tonnes). Prior to 1910, the crops were stored in barns or root cellars, but, by the 1920s, potato cellars or barns came into use. U.S. potato production has increased steadily; two-thirds of the crop comes from Idaho, Washington, Oregon, Colorado, and Maine, and potato growers have strengthened their position in both domestic and foreign markets.

By the 1960s, the Canadian Potato Research Centre in Fredericton, New Brunswick, was one of the top six potato research institutes in the world. Established in 1912 as a dominion experimental station, the station began in the 1930s to concentrate on breeding new varieties of disease-resistant potatoes. In the 1950s–1960s, the growth of the French fry industry in New Brunswick led to a focus on developing varieties for the industry. By the 1970s, the station's potato research was broader than ever before, but the station and its research programs had changed, as emphasis was placed on serving industry rather than potato farmers in general. Scientists at the station even began describing their work using engineering language rather than scientific prose. Potatoes are Canada's most important vegetable crop; they are grown commercially in all its provinces, led by Prince Edward Island. In modern days potatoes have grown in popularity due to their versatility and ability to be used for many different dishes of food.

The first potatoes to arrive in North America from Europe were sent to Gov. Francis Wyatt of Virginia at Jamestown by Nathaniel Butler. There were two large cedar boxes that were filled with different vegetables, among them were potatoes. The very first potatoes that were planted in North America was in Derry New Hampshire in the year 1719. Idaho didn't begin growing potatoes until 1836.

ROLE IN WORLD FOOD SUPPLY

The Food and Agriculture Organization of the United Nations reports that the world production of potatoes in 2013 was about 368 million tonnes. Just over two thirds of the global production is eaten directly by humans with the rest being fed to animals or used to produce starch. This means that the annual diet of an average global citizen in the first decade of the 21st century included about 33 kg (or 73 lb) of potato. However, the local importance of potato is extremely variable and rapidly changing. It remains an essential crop in Europe (especially eastern and central Europe), where per capita production is still the highest in the world, but the most rapid expansion over the past few decades has occurred in southern and eastern Asia. As of 2007, China led the world in potato production, and nearly a third of the world's potatoes were harvested in China and India. The geographic shift of potato production has been away from wealthier countries toward lower-income areas of the world, although the degree of this trend is ambiguous.

In 2008, several international organizations highlighted the potato's role in world food production, in the face of developing economic problems. They cited its potential derived from its status as a cheap and plentiful crop that grows in a wide variety of climates and locales. Due to perishability, only about 5% of the world's potato crop is traded internationally; its minimal presence in world financial markets contributed to its stable pricing during the 2007–2008 world food price crisis. Thus, the United Nations officially declared 2008 as the *International Year of the Potato*, to raise its profile in developing nations, calling the crop a "hidden treasure". This followed the International Rice Year in 2004.

NUTRITION

The potato contains vitamins and minerals, as well as an assortment of phytochemicals, such as carotenoids and natural phenols. Chlorogenic acid constitutes up to 90% of the potato tuber natural phenols. Others found in potatoes are 4-O-caffeoylquinic acid (crypto-chlorogenic acid), 5-O-caffeoylquinic (neo-chlorogenic acid), 3,4-dicaffeoylquinic and 3,5-dicaffeoylquinic acids. A medium-size 150 g (5.3 oz) potato with the skin provides 27 mg of vitamin C (45% of the Daily Value (DV)), 620 mg of potassium (18% of DV), 0.2 mg vitamin B_6 (10% of DV) and trace amounts of thiamin, riboflavin, folate, niacin, magnesium, phosphorus, iron, and zinc.

The potato is best known for its carbohydrate content (approximately 26 grams in a medium potato). The predominant form of this carbohydrate is starch. A small but significant portion of this starch is resistant to digestion by enzymes in the stomach and small intestine, and so reaches the large intestine essentially intact.

This resistant starch is considered to have similar physiological effects and health benefits as fiber: It provides bulk, offers protection against colon cancer, improves glucose tolerance and insulin sensitivity, lowers plasma cholesterol and triglyceride concentrations, increases satiety, and possibly even reduces fat storage. The amount of resistant starch in potatoes depends much on preparation methods. Cooking and then cooling potatoes significantly increases resistant starch. For example, cooked potato starch contains about 7% resistant starch, which increases to about 13% upon cooling.

The storage and cooking method used can significantly affect the nutrient availability of the potato.

Potatoes are often broadly classified as high on the glycemic index (GI) and so are often excluded from the diets of individuals trying to follow a low-GI diet. In fact, the GI of potatoes can vary considerably depending on type (such as red, russet, white, or King Edward), origin (where it was grown), preparation methods (i.e., cooking method, whether it is eaten hot or cold, whether it is mashed or cubed or consumed whole, etc.), and with what it is consumed (i.e., the addition of various high-fat or high-protein toppings).

In the UK, potatoes are not considered by the NHS as counting towards the recommended daily five portions of fruit and vegetables.

COMPARISON TO OTHER MAJOR STAPLE FOODS

The following table shows the nutrient content of potato and other major staple foods, each in respective raw form. Staple foods are not commonly eaten raw and are usually sprouted or cooked before eating. In sprouted and cooked form, the relative nutritional and anti-nutritional contents of each of these grains may be different from the values reported in this table.

Table. Nutrient content of major staple foods

Staple:	RDA	Maize / Corn	Rice (white)	Rice (brown)	Wheat	Potato	Cassava	Soybean (Green)	Sweet potato	Sorghum	Yam	Plantain
Component (per 100g portion)	Amount	Amount	Amount	Amount	Amount	Amount	Amount	Amount	Amount	Amount	Amount	Amount
Water (g)	3000	10	12	10	13	79	60	68	77	9	70	65
Energy (kJ)		1528	1528	1549	1369	322	670	615	360	1419	494	511
Protein (g)	50	9.4	7.1	7.9	12.6	2.0	1.4	13.0	1.6	11.3	1.5	1.3
Fat (g)		4.74	0.66	2.92	1.54	0.09	0.28	6.8	0.05	3.3	0.17	0.37
Carbohydrates (g)	130	74	80	77	71	17	38	11	20	75	28	32
Fiber (g)	30	7.3	1.3	3.5	12.2	2.2	1.8	4.2	3	6.3	4.1	2.3
Sugar (g)		0.64	0.12	0.85	0.41	0.78	1.7	0	4.18	0	0.5	15
Calcium (mg)	1000	7	28	23	29	12	16	197	30	28	17	3
Iron (mg)	8	2.71	0.8	1.47	3.19	0.78	0.27	3.55	0.61	4.4	0.54	0.6
Magnesium (mg)	400	127	25	143	126	23	21	65	25	0	21	37
Phosphorus (mg)	700	210	115	333	288	57	27	194	47	287	55	34
Potassium (mg)	4700	287	115	223	363	421	271	620	337	350	816	499
Sodium (mg)	1500	35	5	7	2	6	14	15	55	6	9	4
Zinc (mg)	11	2.21	1.09	2.02	2.65	0.29	0.34	0.99	0.3	0	0.24	0.14
Copper (mg)	0.9	0.31	0.22		0.43	0.11	0.10	0.13	0.15	-	0.18	0.08
Manganese (mg)	2.3	0.49	1.09	3.74	3.99	0.15	0.38	0.55	0.26	-	0.40	-
Selenium (μg)	55	15.5	15.1		70.7	0.3	0.7	1.5	0.6	0	0.7	1.5
Vitamin C (mg)	90	0	0	0	0	19.7	20.6	29	2.4	0	17.1	18.4
Thiamin (B1)(mg)	1.2	0.39	0.07	0.40	0.30	0.08	0.09	0.44	0.08	0.24	0.11	0.05
Riboflavin (B2)(mg)	1.3	0.20	0.05	0.09	0.12	0.03	0.05	0.18	0.06	0.14	0.03	0.05
Niacin (B3) (mg)	16	3.63	1.6	5.09	5.46	1.05	0.85	1.65	0.56	2.93	0.55	0.69
Pantothenic acid (B5) (mg)	5	0.42	1.01	1.49	0.95	0.30	0.11	0.15	0.80	-	0.31	0.26
Vitamin B6 (mg)	1.3	0.62	0.16	0.51	0.3	0.30	0.09	0.07	0.21	-	0.29	0.30

Folate Total (B9) (µg)	400	19	8	20	38	16	27	165	11	0	23	22
Vitamin A (IU)	5000	214	0	0	9	2	13	180	14187	0	138	1127
Vitamin E, alpha-tocopherol (mg)	15	0.49	0.11	0.59	1.01	0.01	0.19	0	0.26	0	0.39	0.14
Vitamin K1 (µg)	120	0.3	0.1	1.9	1.9	1.9	1.9	0	1.8	0	2.6	0.7
Beta-carotene (µg)	10500	97	0		5	1	8	0	8509	0	83	457
Lutein+zeaxanthin (µg)		1355	0		220	8	0	0	0	0	0	30
Saturated fatty acids (g)		0.67	0.18	0.58	0.26	0.03	0.07	0.79	0.02	0.46	0.04	0.14
Monounsaturated fatty acids (g)		1.25	0.21	1.05	0.2	0.00	0.08	1.28	0.00	0.99	0.01	0.03
Polyunsaturated fatty acids (g)		2.16	0.18	1.04	0.63	0.04	0.05	3.20	0.01	1.37	0.08	0.07

Corn, yellow	Rice, white, long-grain, regular, raw, unenriched
Wheat, hard red winter	Potato, flesh and skin, raw
Cassava, raw	Soybeans, green, raw
Sweet potato, raw, unprepared	Sorghum, raw
Yam, raw	Plantains, raw
Rice, brown, long-grain, raw	

TOXICITY

Potatoes contain toxic compounds known as glycoalkaloids, of which the most prevalent are solanine and chaconine. Solanine is also found in other plants in the family Solanaceae, which includes such plants as the deadly nightshade (*Atropa belladonna*), henbane (*Hyoscyamus niger*), tobacco (*Nicotiana*), as well as eggplant and tomato.

These compounds, which protect the plant from its predators, are, in general, concentrated in its leaves, stems, sprouts, and fruits. In a summary of several studies, the glycoalkaloid content was highest in flowers and sprouts and lowest in the tuber flesh (in order from highest to lowest content, generally: flowers, sprouts, leaves, skin, roots, berries, peel [skin plus outer cortex of tuber flesh], stems, and tuber flesh). Exposure to light, physical damage, and age increase glycoalkaloid content within the tuber. Cooking at high temperatures —over 170 °C (338 °F) — partly destroys these. The concentration of glycoalkaloid in wild potatoes suffices to produce toxic effects in humans. Glycoalkaloids may cause headaches, diarrhea, cramps, and in severe cases coma and death; however, poisoning from potatoes occurs very rarely. Light exposure causes greening from chlorophyll synthesis, thus giving a visual clue as to areas of the tuber that may have become more toxic; however, this does not provide a definitive guide, as greening and glycoalkaloid accumulation can occur independently of each other. Some varieties of potato contain greater glycoalkaloid concentrations than others; breeders developing new varieties test for this, and sometimes have to discard an otherwise promising cultivar.

Breeders try to keep solanine levels below 200 mg/kg (200 ppmw). However, when these commercial varieties turn green, even they can approach concentrations of solanine of 1000 mg/kg (1000 ppmw). In normal potatoes, analysis has shown solanine levels may be as little as 3.5% of the breeders' maximum, with 7–187 mg/kg being found. While a normal potato has 12–20 mg/kg of glycoalkaloid content, a green tuber contains 250–280 mg/kg, and green skin 1500–2200 mg/kg.

The U.S. National Toxicology Program suggests that the average American consume at most 12.5 mg/day of solanine from potatoes (the toxic dose is actually several times this, depending on body weight). Douglas L. Holt, the State Extension Specialist for Food Safety at the University of Missouri, notes that no reported cases of potato-source solanine poisoning have occurred in the U.S. in the last 50 years, and most cases involved eating green potatoes or drinking potato-leaf tea.

GROWTH AND CULTIVATION

Potatoes are generally grown from seed potatoes – these are tubers specifically grown to be disease free and provide consistent and healthy plants. To be disease free, the areas where seed potatoes are grown are selected with care. In the USA this restricts production of seed potatoes to only 15 states out of the 50 states that grow potatoes. These locations are selected for their cold hard winters that kill pests and long sunshine hours in the summer for optimum growth. In the UK, most seed potatoes originate in Scotland in areas where westerly winds prevent aphid attack and thus prevent spread of potato virus pathogens. Potato growth has been divided into five phases. During the first phase, sprouts emerge from the seed potatoes and root growth begins. During the second, photosynthesis begins as the plant develops leaves and branches. In the third phase stolons develop from lower leaf axils on the stem and grow downwards into the ground and on these stolons new tubers develop as swellings of the stolon. This phase is often (but not always) associated with flowering. Tuber formation halts when soil temperatures reach 27 °C (81 °F); hence potatoes are considered a cool-season crop. Tuber bulking occurs during the fourth phase, when the plant begins investing the majority of its resources in its newly formed tubers. At this stage, several factors are critical to yield: optimal soil moisture and temperature, soil nutrient availability and balance, and resistance to pest attacks. The final phase is maturation: The plant canopy dies back, the tuber skins harden, and their sugars convert to starches.

New tubers may arise at the soil surface. Since exposure to light leads to greening of the skins and the development of solanine, growers are interested in covering such tubers. Commercial growers usually address this problem by piling additional soil around the base of the plant as it grows ("hilling", or in British English "earthing up"). An alternative method used by home gardeners and smaller-scale growers involves covering the growing area with organic mulches such as straw or with plastic sheets.

Correct potato husbandry can be an arduous task in some circumstances. Good ground preparation, harrowing, plowing, and rolling are always needed, along with a little grace from the weather and a good source of water. Three successive plowings, with associated harrowing and rolling, are desirable before planting. Eliminating all root-weeds is desirable in potato cultivation. In general, the potatoes themselves are grown from the eyes of another potato and not from seed. Home gardeners often plant a piece of potato with two or three eyes in a hill of mounded soil. Commercial growers plant potatoes as a row crop using seed tubers, young plants or microtubers and may mound the entire row. Seed potato crops are 'rogued' in some countries to eliminate diseased plants or those of a different variety from the seed crop. Potatoes are sensitive to heavy frosts, which damage them in the ground. Even cold weather makes potatoes more susceptible to bruising and possibly later rotting, which can quickly ruin a large stored crop.

At harvest time, gardeners usually dig up potatoes with a long-handled, three-prong "grape" (or graip), i.e., a spading fork, or a potato hook, which is similar to the graip but with tines at a 90° angle to the handle. In larger plots, the plow is the fastest implement for unearthing potatoes. Commercial harvesting is typically done with large potato harvesters, which scoop up the plant and surrounding earth. This is transported up an apron chain consisting of steel links several feet wide, which separates some of the dirt. The chain deposits into an area where further separation occurs. Different designs use different systems at this point. The most complex designs use vine choppers and shakers, along with a blower system to separate the potatoes from the plant. The result is then usually run past workers who continue to sort out plant material, stones, and rotten potatoes before the potatoes are continuously delivered to a wagon or truck. Further inspection and separation occurs when the potatoes are unloaded from the field vehicles and put into storage.

Immature potatoes may be sold as "new potatoes" and are particularly valued for taste. These are often harvested by the home gardener or farmer by "grabbling", i.e. pulling out the young tubers by hand while leaving the plant in place. Potatoes are usually cured after harvest to improve skin-set. Skin-set is the process by which the skin of the potato becomes resistant to skinning damage. Potato tubers may be susceptible to skinning at harvest and suffer skinning damage during harvest and handling operations. Curing allows the skin to fully set and any wounds to heal. Wound-healing prevents infection and water-loss from the tubers during storage. Curing is normally done at relatively warm temperatures 50 to 60 °C (122 to 140 °F) with high humidity and good gas-exchange if at all possible.

STORAGE

Storage facilities need to be carefully designed to keep the potatoes alive and slow the natural process of decomposition, which involves the breakdown of starch. It is crucial that the storage area is dark, well ventilated and for long-term storage maintained at temperatures near 4 °C (39 °F). For short-term storage before cooking, temperatures of about 7 to 10 °C (45 to 50 °F) are preferred. On the other hand, temperatures below 4 °C (39 °F) convert potatoes' starch into sugar, which alters their taste and cooking qualities and leads to higher acrylamide levels in the cooked product, especially in deep-fried dishes—the discovery of acrylamides in starchy foods in 2002 has led to many international health concerns as they are believed to be possible carcinogens and their occurrence in cooked foods is currently under study as a possible influence in potential health problems.

Under optimum conditions possible in commercial warehouses, potatoes can be stored for up to ten to twelve months. When stored in homes, the shelf life is usually only a few weeks. If potatoes develop green areas or start to sprout, these areas should be trimmed before using. Trimming or peeling green areas are inadequate to remove copresent toxins, and such potatoes are no longer suitable as animal food. Commercial storage of potatoes involves several phases: drying of surface moisture; a wound healing phase at 85% to 95% relative humidity and temperatures below 25 °C (77 °F); a staged cooling phase; a holding phase; and a reconditioning phase, during which the tubers are slowly warmed. Mechanical ventilation is used at various points during the process to prevent condensation and accumulation of carbon dioxide.

YIELD

The world dedicated 18.6 million hectares in 2010 for potato cultivation. The average world farm yield for potato was 17.4 tonnes per hectare, in 2010. Potato farms in the United States were the most productive in 2010, with a nationwide average of 44.3 tonnes per hectare. United Kingdom was a close second. New Zealand farmers have demonstrated some of the best commercial yields in the world, ranging between 60 to 80 tonnes per hectare, some reporting yields of 88 tonnes potatoes per hectare.

There is a big gap among various countries between high and low yields, even with the same variety of potato. Average potato yields in developed economies ranges between 38–44 tonnes per hectare. China and India accounted for over a third of world's production in 2010, and had yields of 14.7 and 19.9 tonnes per hectare respectively. The yield gap between farms in developing economies and developed economies represents an opportunity loss of over 400 million tonnes of potato, or an amount greater than 2010 world potato production. Potato crop yields are determined by factors such as the crop breed, seed age and quality, crop management practices and the plant environment. Improvements in one or more of these yield determinants, and a closure of the yield gap, can be a major boost to food supply and farmer incomes in the developing world.

VARIETIES

While there are close to 4000 different varieties of potato, it has been bred into many standard or well-known varieties, each of which has particular agricultural or culinary attributes. In general, varieties are categorized into a few main groups, such as russets, reds, whites, yellows (also called Yukons) and purples—based on common characteristics. Around 80 varieties are commercially available in the UK. For culinary purposes, varieties are often differentiated by their waxiness. Floury, or mealy (baking) potatoes have more starch (20–22%) than waxy (boiling) potatoes (16–18%). The distinction may also arise from variation in the comparative ratio of two potato starch compounds: amylose and amylopectin. Amylose, a long-chain molecule, diffuses from the starch granule when cooked in water, and lends itself to dishes where the potato is mashed. Varieties that contain a slightly higher amylopectin content, a highly branched molecule, help the potato retain its shape when boiled. The European Cultivated Potato Database (ECPD) is an online collaborative database of potato variety descriptions, updated and maintained by the Scottish Agricultural Science Agency within the framework of the European Cooperative Programme for Crop Genetic Resources Networks (ECP/GR)—which is run by the International Plant Genetic Resources Institute (IPGRI).

Popular varieties (cultivars) include:

- Adirondack Blue
- Adirondack Red
- Agata
- Almond
- Alpine Russet
- Alturas
- Amandine
- Annabelle
- Anya
- Arran Victory
- Atlantic
- Austrian Crescent
- Avalanche
- Bamberg
- Bannock Russet
- Belle de Fontenay
- BF-15
- Bildtstar
- Bintje
- Blazer Russet
- Blue Congo
- Bonnotte
- British Queens
- Cabritas
- Camota
- Canela Russet
- Cara
- Carola
- Chelina
- Chiloé

- Cielo
- Clavela Blanca
- Désirée
- Estima
- Fianna
- Fingerling
- Flava
- French Fingerling
- German Butterball
- Golden Wonder
- Goldrush
- Home Guard
- Innovator
- Irish Cobbler
- Irish Lumper
- Jersey Royal
- Kennebec
- Kerr's Pink
- Kestrel
- Keuka Gold
- King Edward
- Kipfler
- Lady Balfour
- Langlade
- Linda potato
- Marcy
- Marfona
- Maris Piper
- Marquis
- Megachip
- Melody
- Monalisa
- Nicola
- Norgold Russet
- Pachacoña
- Pike
- Pink Eye
- Pink Fir Apple
- Primura
- Ranger Russet
- Ratte
- Record

- Red La Soda
- Red Norland
- Red Pontiac
- Rooster
- Russet Burbank
- Russet Norkotah
- Selma
- Shepody
- Sieglinde
- Silverton Russet
- Sirco
- Snowden
- Spunta
- Up to date
- Stobrawa
- Superior
- Umatilla Russet
- Villetta Rose
- Vivaldi
- Vitelotte
- Yellow Finn
- Yukon Gold

BLUE VARIETIES

The blue or purple potato originated in South America. It has purple skin and flesh, which becomes blue once cooked. It has a slight whitish scab that seems to be present in all samples. The variety, called "Cream of the Crop", has been introduced into Ireland and has proved popular. A mutation in the varieties' P locus causes production of the antioxidant anthocyanin.

GENETICALLY MODIFIED POTATOES

A genetically modified potato is a potato that has had its genes modified, using genetic engineering. Goals of modification include introducing pest resistance, tweaking the amounts of certain chemicals produced by the plant, and to prevent browning or bruising of the tubers. Varieties modified to produce large amounts of starches may be approved for industrial use only, not for food.

CURRENTLY MARKETED VARIETIES

Used for food

Innate

The genetically modified Innate potato was approved by the USDA in 2014 and the FDA in 2015. The cultivar was developed by J. R. Simplot Company. It is designed to resist blackspot bruising, browning and to contain less of the amino acid asparagine that turns into acrylamide during the frying of potatoes. Acrylamide is a probable human carcinogen, so reduced levels of it in fried potato foods is desirable. The 'Innate' name comes from the fact that this variety does not contain any genetic material from other species (the genes used are "innate" to potatoes) and uses RNA interference to switch off genes. Simplot hopes that not including genes from other species will assuage consumer fears about biotechnology.

The "Innate" potato is not a single cultivar, rather, it is a group of potato varieties that have had the same genetic alterations applied using the same process. Five different potato varieties have been transformed, creating "innate"versions of the varieties, with all of the original traits, plus the engineered ones.

Ranger Russet, Russet Burbank,and Atlantic potatoes have all been transformed by Simplot, as well as and two proprietary varieties. Modifications of each variety involved two transformations, one for each of the two new traits, thus there was a total of ten transformation events in developing the different Innate varieties.

McDonald's is a major consumer of potatoes in the US, but the Food and Water Watch has petitioned them to reject the newly marketed Innate potatoes. McDonald's has announced that they have ruled out using Innate.

PREVIOUSLY MARKETED VARIETIES

Used for food

New Leaf

In 1995, Monsanto introduced the NewLeaf variety of potato which was their first genetically modified crop. It was designed to resist attack from the Colorado potato beetle due to the insertion of Bt toxin producing genes from the bacterium *Bacillus thuringiensis*. The insect-resistant potatoes found only a small market, and Monsanto discontinued the sale of seed in 2001.

Used in industry

Amflora

'Amflora' (also known as EH92-527-1) was a cultivar developed by BASF Plant Science for production of pure amylopectin starch for processing into waxy potato starch.

It was approved for industrial applications in the European Union market on 2 March 2010 by the European Commission, but was withdrawn from the EU market in January 2012 due to a lack of acceptance from farmers and consumers.

UNMARKETED VARIETIES

A modified Désirée potato was developed in the 1990s by biochemist John Gatehouse at Cambridge Agricultural Genetics (later renamed Axis Genetics) and had gone through two years of field trials at Rothamsted Experimental Station. The potatoes were modified to express the *Galanthus nivalis agglutinin* (GNA) gene from the *Galanthus* (snowdrop) plant, which caused them to produce GNA lectin protein that is toxic to some insects. This variety of potatoes is the one which was involved in the Pusztai affair.

In 2014, a team of British scientists published a paper about three-year field trial showing that another genetically modified version of the Désirée cultivar can resist infection after exposure to late blight, one of the most serious diseases of potatoes. They developed this potato for blight resistance by inserting a gene (Rpi-vnt1.1), into the DNA of Désirée potatoes. This gene, which conferred the resistance to blight, was isolated from a wild relative of potatoes, *Solanum venturii*, which is a native of South America.

USES

- Potatoes are used to brew alcoholic beverages such as vodka, potcheen, or akvavit.
- They are also used as food for domestic animals.
- Potato starch is used in the food industry as, for example, thickeners and binders of soups and sauces, in the textile industry, as adhesives, and for the manufacturing of papers and boards.
- Maine companies are exploring the possibilities of using waste potatoes to obtain polylactic acid for use in plastic products; other research projects seek ways to use the starch as a base for biodegradable packaging.
- Potato skins, along with honey, are a folk remedy for burns in India. Burn centers in India have experimented with the use of the thin outer skin layer to protect burns while healing.
- Potatoes (mainly Russets) are commonly used in plant research. The consistent parenchyma tissue, the clonal nature of the plant and the low metabolic activity provide a very nice "model tissue" for experimentation. Wound-response studies are often done on potato tuber tissue, as are electron transport experiments. In this respect, potato tuber tissue is similar to *Drosophila melanogaster*, *Caenorhabditis elegans* and *Escherichia coli*: they are all "standard" research organisms.

CULINARY USES

Potatoes are prepared in many ways: skin-on or peeled, whole or cut up, with seasonings or without. The only requirement involves cooking to swell the starch granules. Most potato dishes are served hot, but some are first cooked, then served cold, notably potato salad and potato chips/crisps. Common dishes are: mashed potatoes, which are first boiled (usually peeled), and then mashed with milk or yogurt and butter; whole baked potatoes; boiled or steamed potatoes; French-fried potatoes or chips; cut into cubes and roasted; scalloped, diced, or sliced and fried (home fries); grated into small thin strips and fried (hash browns); grated and formed into dumplings, Rösti or potato pancakes.

Unlike many foods, potatoes can also be easily cooked in a microwave oven and still retain nearly all of their nutritional value, provided they are covered in ventilated plastic wrap to prevent moisture from escaping; this method produces a meal very similar to a steamed potato, while retaining the appearance of a conventionally baked potato. Potato chunks also commonly appear as a stew ingredient. Potatoes are boiled between 10 and 25 minutes, depending on size and type, to become soft.

Grading

In the U.S., potato grading for Idaho potatoes is performed in which No. 1 potatoes are the highest quality and No. 2 are rated as lower in quality due to their appearance (e.g. blemishes or bruises, pointy ends). Potato density assessment can be performed by floating them in brines. High-density potatoes are desirable in the production of dehydrated mashed potatoes, potato crisps and french fries.

Latin America

Peruvian cuisine naturally contains the potato as a primary ingredient in many dishes, as around 3,000 varieties of this tuber are grown there. Some of the more notable dishes include boiled potato as a base for several dishes or with ají-based sauces like in Papa a la Huancaína or ocopa, diced potato for its use in soups like in cau cau, or in Carapulca with dried potato (papa seca). Smashed condimented potato is used in causa Limeña and papa rellena. French-fried potatoes are a typical ingredient in Peruvian stir-fries, including the classic dish lomo saltado.

Chuño is a freeze-dried potato product traditionally made by Quechua and Aymara communities of Peru and Bolivia, and is known in various countries of South America, including Peru, Bolivia, Argentina, and Chile. In Chile's Chiloé Archipelago, potatoes are the main ingredient of many dishes, including milcaos, chapaleles, curanto and chochoca. In Ecuador, the potato, as well as being a staple with most dishes, is featured in the hearty *locro de papas*, a thick soup of potato, squash, and cheese.

European cuisine

In the UK, potatoes form part of the traditional staple fish and chips. Roast potatoes are commonly served with a Sunday roast, and mashed potatoes form a major component of several other traditional dishes such as shepherd's pie, bubble and squeak, and bangers and mash. New potatoes are often cooked with mint and served with a little melted butter. The Tattie scone is a popular Scottish dish containing potatoes. Colcannon is a traditional Irish food made with mashed potato, shredded kale or cabbage, and onion; champ is a similar dish. Boxty pancakes are eaten throughout Ireland, although associated especially with the north, and in Irish diaspora communities; they are traditionally made with grated potatoes, soaked to loosen the starch and mixed with flour, buttermilk and baking powder. A variant eaten and sold in Lancashire, especially Liverpool, is made with cooked and mashed potatoes.

Bryndzové halušky is the Slovakian national dish, made of a batter of flour and finely grated potatoes that is boiled to form dumplings. These are then mixed with regionally varying ingredients. In Northern and Eastern Europe, especially in Scandinavian countries, Poland, Russia, Belarus and Ukraine, newly harvested, early ripening varieties are considered a special delicacy. Boiled whole and served un-peeled with dill, these "new potatoes" are traditionally consumed with Baltic herring. Puddings made from grated potatoes (kugel, kugelis, and potato babka) are popular items of Ashkenazi, Lithuanian, and Belarusian cuisine.

Cepelinai is Lithuanian national dish. They are a type of dumpling made from riced potatoes (see Potato ricer) and usually stuffed with minced meat, although sometimes dry cottage cheese (curd) or mushrooms are used instead. In Western Europe, especially in Belgium, sliced potatoes are fried to create *frieten*, the original French fried potatoes. *Stamppot*, a traditional Dutch meal, is based on mashed potatoes mixed with vegetables. In France, the most notable potato dish is the *Hachis Parmentier*, named after Antoine-Augustin Parmentier, a French pharmacist, nutritionist, and agronomist who, in the late 18th century, was instrumental in the acceptance of the potato as an edible crop in the country. The *pâté aux pommes de terre* is a regional potato dish from the central Allier and Limousin regions.

In the north of Italy, in particular, in the Friuli region of the northeast, potatoes serve to make a type of pasta called gnocchi. Similarly, cooked and mashed potatoes or potato flour can be used in the Knödel or dumpling eaten with or added to meat dishes all over central and Eastern Europe, but especially in Bavaria and Luxembourg. Potatoes form one of the main ingredients in many soups such as the vichyssoise and Albanian potato and cabbage soup. In western Norway, komle is popular. A traditional Canary Islands dish is Canarian wrinkly potatoes or *papas arrugadas*. *Tortilla de patatas* (potato omelete) and *patatas bravas* (a dish of fried potatoes in a spicy tomato sauce) are near-universal constituent of Spanish tapas.

North America

In the United States, potatoes have become one of the most widely consumed crops and thus have a variety of preparation methods and condiments.

French fries and often hash browns are commonly found in typical American fast-food burger joints and cafeterias. One popular favorite involves a baked potato with cheddar cheese (or sour cream and chives) on top, and in New England "smashed potatoes" (a chunkier variation on mashed potatoes, retaining the peel) have great popularity. Potato flakes are popular as an instant variety of mashed potatoes, which reconstitute into mashed potatoes by adding water, with butter or oil and salt to taste.

A regional dish of Central New York, salt potatoes are bite-size new potatoes boiled in water saturated with salt then served with melted butter. At more formal dinners, a common practice includes taking small red potatoes, slicing them, and roasting them in an iron skillet. Among American Jews, the practice of eating latkes (fried potato pancakes) is common during the festival of Hanukkah.

A traditional Acadian dish from New Brunswick is known as *poutine râpée*. The Acadian poutine is a ball of grated and mashed potato, salted, sometimes filled with pork in the center, and boiled. The result is a moist ball about the size of a baseball. It is commonly eaten with salt and pepper or brown sugar. It is believed to have originated from the German *Klöße*, prepared by early German settlers who lived among the Acadians. *Poutine*, by contrast, is a hearty serving of French fries, fresh cheese curds and hot gravy. Tracing its origins to Quebec in the 1950s, it has become a widespread and popular dish throughout Canada.

South Asia

In South Asia, Potato is very popular traditional staple. In India, the most popular potato dishes are *aloo ki sabzi*, batata vada, and samosa, which is spicy mashed potato mixed with a small amount of vegetable stuffed in conical dough, and deep fried. Potatoes are also a major ingredient as fast food items, such as aloo chaat, where they are deep fried and served with chutney. In Northern India, alu dum and alu paratha are a favorite part of the diet; the first is a spicy curry of boiled potato, the second is a type of stuffed chapati.

A dish called masala dosa from South India is very notable all over India. It is a thin pancake of rice and pulse paste rolled over spicy smashed potato and eaten with sambhar and chutney. Poori in south India in particular in Tamil Nadu is almost always taken with smashed potato masal. Other favorite dishes are alu tikki and pakoda items. Vada pav is a popular vegetarian fast food dish in Mumbai and other regions in the Maharashtra in India.

Aloo posto (a curry with potatoes and poppy seeds) is immensely popular in East India, especially Bengal. Although potatoes are not native to India, it has become a vital part of food all over the country especially North Indian food preparations. In Tamil Nadu this tuber acquired a name based on its appearance 'urulai-k-kizhangu' meaning cylindrical tuber. The Aloo gosht, Potato and meat curry, is one of the popular dishes in South Asia, especially in Pakistan.

East Asia

In East Asia, particularly Southeast Asia, rice is by far the predominant starch crop, with potatoes a secondary crop, especially in China and Japan. However, it is used in northern China where rice is not easily grown, with a popular dish being, made with green pepper, vinegar and thin slices of potato. In the winter, roadside sellers in northern China will also sell roasted potatoes. It is also occasionally seen in Korean and Thai cuisines.

ART

The potato has been an essential crop in the Andes since the pre-Columbian Era. The Moche culture from Northern Peru made ceramics from earth, water, and fire. This pottery was a sacred substance, formed in significant shapes and used to represent important themes. Potatoes are represented anthropomorphically as well as naturally. During the late 19th century, numerous images of potato harvesting appeared in European art, including the works of Willem Witsen and Anton Mauve. Van Gogh's 1885 painting *The Potato Eaters* portrays a family eating potatoes. Invented in 1949 and marketed and sold commercially by Hasbro in 1952, Mr. Potato Head is an American toy that consists of a plastic potato and attachable plastic parts such as ears and eyes to make a face. It was the first toy ever advertised on television.

3

Potato Breeding

POTATO INTRODUCTIONS AND BREEDING UP TO THE EARLY 20TH CENTURY

Modern potato varieties usually have in their ancestries primitive cultivars or wild species introduced during the present century and may superficially appear to be of very diverse origins. However the breeding procedures used have necessarily involved repeated back-crossing to existing stocks to eliminate the 'primitive' characters of the introductions, and generally 80 % or more of the genes of a variety trace to varieties grown early in this century.

THE LATIN AMERICAN GENE-POOL

Cultivated potatoes

Although 'primitive' cultivated potatoes now occur in central America it is believed that prior to the Spanish Conquest they were restricted to South America where they occur at high altitudes in the Andes from Colombia and Venezuela through Ecuador, Peru and Bolivia to northern Argentina, with an outlying population further south at low altitude in Chile. Their climatic range is from near-alpine to temperate but, apart from the Chilean outlier at a latitude about equivalent to Spain, they occur in tropical latitudes where the days are always short. There are a large number of varieties. They are mainly tetraploid but there is a fair representation of diploids, triploids and pentaploids also, although the Chilean population appears to be wholly tetraploid.

Current practise is to treat them all, apart from a very few known to be hybrids involving wild species, as belonging to the species *Solanum tuberosum* then to subdivide that species either in terms of subspecies, using a classical taxonomic system, or in terms of groups using the horticultural system. In both cases the primary separation is on chromosome number. Variation within each ploidy level is wide, especially so in the tetraploids, and the ranges of variation overlap. It is rarely possible to be sure which subspecies or group a cultivar belongs to without counting the chromosomes. Two main groups are recognized within the diploids. The main distinction between them, which appears rather arbitrary, is that the tubers of group *Phureja* are less dormant.

This group tends to have a more northern distribution within South America and to be grown at lower altitudes where warmer conditions permit replanting soon after harvest to secure more than one crop in the year. The more dormant group *Stenotomum* mainly occurs further south. One subgroup within *Stenotomum,* subgroup *Goniocalyx,* includes the ' Golden Potato of Peru' which has long been noted for its outstanding culinary quality. Another diploid type, *Ajanhuiri,* is frost resistant and is grown at high altitudes but has a bitter tuber which requires special treatment before use; it has recently been determined to be of hybrid origin involving the wild species *S. megistacrolobum.*

Two general types of triploids are recognized. Group *Chaucha* appears to derive from crossing of diploid and tetraploid cultivars while the other type, which has similar properties to *Ajanhuiri* and has long been recognized as being derived from hybridization involving the wild species *S. acaule*\ is treated as a separate species, *S. xjuzepczukii.* Triploids are effectively sterile and are of little practical interest as parents for breeding. The tetraploids, apart from the Chileans, are classed as group *Andigena* and occur throughout the Andean range outlined above.

The diploids are obviously ancestral to them but the tetraploids do not appear simply to be doubled-up diploids; they have some characters which are not found in the diploids and hybridization with wild species is probably involved in their ancestry. The Chilean tetraploids are adapted to a climate more nearly resembling Britain as regards day lengths, and tend to have larger leaves than the Andean potatoes. At one time, before anything was known of chromosome numbers, it was thought that they might have been the original potatoes from which the Andean population was derived. Wight argued against this view on the basis of the greater numbers of cultivars and the greater range of variation between them in the Andes than in Chile and also on other grounds, such as the languages in which the varieties were named, the fact that high-quality potatoes introduced to Chile from Peru would not yield sufficiently to be maintained in cultivation thus being unlikely to be re-introductions and that, prior to the Conquest, agriculture was more developed in the Andes than in Chile where the native peoples could subsist to a large extent by hunting.

The suggestion that the Chilean potatoes are the originals of all potatoes is no longer maintained but there are still some who view them, because of their supposed nearer resemblance to European potatoes, as being the origin of our potatoes and who also argue for their having evolved independently from wild species in Chile rather than being derived from the Andean population. However, the historical evidence suggests that it is unlikely that Chilean potatoes were the first to be introduced to Europe though some subsequent introductions may have that origin. It also seems that the differences from the

Andean potatoes may have been over-stressed. A better adaptation to the British climate than that of the Chileans has been achieved in the development of Neo-Tuberosum potatoes from *Andigena* and a study of Chilean potatoes led to the conclusion that they probably represent a small and highly inbred sub-population derived from Andigena.

Also, there are few wild species in Chile and many in the Andes. The pentaploids are apparently derived from hybridization involving the triploid *S. xjuzepczukii* and are relatively sterile and of little apparent interest as parents for breeding. They are termed *S.* x *curtilobum.* Until recently, cultivars grown by the native peoples for their own use were mixed seemingly haphazardly. Wight said ' Very often he will grow a dozen or more in one field, not with each variety in a separate row or portion of the plot, but indiscriminately mixed together... '.

More recently, Jackson, Hawkes and Rowe studied two fields in an area of traditional agriculture finding a total of 25 varieties, three diploid, three triploid and the others tetraploid. This haphazard mixing may suggest that the farmers viewed the varieties as all being essentially alike, simply 'potatoes', but this is far from being the case. Wight says 'many of these varieties will be named and well known in the immediate vicinity; some... may be quite widely known' and he refers to the perplexity of the farmer finding a variety he does not recognize in his field, presumably from a self-sown seedling.

Jackson *et al.* found that the farmers had a name for each variety and knew their culinary qualities and had at least some knowledge of their individual cultural requirements. Eloury varieties were eaten boiled or mashed, intermediate ones boiled or in soups, and watery ones in soups, and in spite of the extremely subjective nature of fiavour assessment there was remarkable agreement between farmers as to which varieties had the best fiavours.

With regard to one superior variety, the farmers said that it was grown only on the terraced valley sides where drainage was good because its yield was poor and its quality depressed when it was grown on the wetter riverside fields. But a mystery remained as to why, in spite of the detailed knowledge of the varieties thus revealed, the most favoured variety of all as regards quality was absent from one field and constituted only four of the 326 plants in the other, and why some varieties of low quality were extensively grown.This system of mixed culture is going out of fashion and, as noted, the fields studied by Jackson *et al.*

were in a relict area of traditional culture. The more usual situation now is mono-culture of modern high-yielding varieties bred within South America. A few such varieties are extensively grown and are rapidly displacing the traditional cultivars. Concern at this situation has led to the establishment of the International Potato Centre in Peru to assemble and conserve the variation before all is lost. The Centre has assembled over 13 000 accessions of primitive cultivars of which about 13, 10, 74 and 3% were diploid, triploid, tetraploid and pentaploid respectively.

These included many duplications, the same varieties obtained from different places perhaps with different names and, following discarding of duplicates, 6000 are now maintained. The variation between native cultivars is greater than between modern varieties. This can be conveniently demonstrated by reference to the most readily observable characters, the shapes of the tubers and the colours of the skin and flesh. In these respects we have fairly definite requirements, a smooth and fairly compact tuber shape easy to peel, and a white or cream flesh though in some countries yellow is preferred.

But various skin colours are acceptable. Most of the variants are undesirable from a modern cultivator's point of view, but from this readily observed variation one may draw inferences regarding other less obvious variation, for example in disease resistances or culinary qualities. Tubers vary in length from being shorter than broad through round and oval to sausage-like shapes several times as long as broad; the longer shapes may be straight, or kidney- or sickle-shaped, or in the extreme, of a coiled form. In cross-section tubers may be round or be flattened, in some cases being more than three times as broad as thick.

They may be rounded or be tapered at the ends, or may tend to be rectangular. Some are broader at the base or ' heel' where they are attached to the stolon, while some are narrower at the heel and broader at the tip. The main growing point or ' rose', around which the terminal eyes are centred, is normally at the tip but sometimes is offset. The eyes may be few but they are often more numerous than in modern varieties; they may be mainly centred around the ' rose' or be widely distributed over the tuber.

They may be shallow, as commonly preferred but sometimes they protrude and frequently they are deeper than would now be tolerated. Deep or protruding eyes, or swellings below eyes, lead to irregularities of tuber shape and in extreme cases to shapes described as resembling concertinas or pineapples. Branching may also occur, in the extreme leading to a shape resembling a hand with fingers. The skins may lack anthocyanin pigmentation, being then white, yellowish or brownish, or pigments ranging from a faint pink through richer pinks and reds to purplish-red, purple, or dark purple-black may occur. These pigments may cover the whole skin or be restricted to the eyes, or eyes and brows, or be predominantly towards the rose end, or may be scattered, stippled or streaked over the tuber; in some cases they are absent round the eyes but present elsewhere on the tuber. The basic range of flesh colours is from white to rich yellow, but when anthocyanin occurs in the skin it often also occurs in the flesh with the same range from pink to dark purple.

When it occurs it may be restricted to the vascular ring, then as scattered spots or larger patches or completely colouring the ring, or it may extend to the pith within the ring either patchily or completely colouring the pith, and it may also extend to the cortex between the vascular ring and the skin. In extreme cases, the whole flesh may be red or purple. There is also variation from bright, glossy to dull, netted or russeted skin, and flesh colours may show through the skin giving, for example, an orange appearance with a pink skin over a yellow flesh. While there is great variation between cultivars in each area and even within single flelds, it does not follow that a sample of the variation from one area will adequately represent the total variation.

There are some geographical trends, with characters which occur throughout the range being more frequent in the northern than the southern part of South America or *vice versa,* and with scarcer characters perhaps having only local distributions. A few primitive cultivars have been introduced to modern breeding. A Chilean tetraploid ' Villaroela' was used in the United States early this century for virus resistance and occurs in the ancestries of a number of varieties, including the British cultivars Pentland Crown and Croft. An *Andigena* from Bolivia, Commonwealth Potato Collection accession (CPC) No. 1673, was used for its eel worm resistance and occurs in many ancestries.

Three accessions of a diploid of the *Phureja* type were used at the Scottish Plant Breeding Station, one in a cross with *S. demissum* and also in a direct cross with a tetraploid cultivar, both hybrids occurring in the ancestries of several varieties; another is in the ancestry of Pentland Glory; the third is in the pedigrees of some varieties released abroad. The Scottish variety Dr Mclntosh also has *rybinii* in its ancestry, and some German breeding lines are believed to involve *Andigena.* However, these accessions contribute little to the total composition of the resultant varieties and the major contribution made by primitive cultivars is from the earlier introductions which gave rise to the potato varieties of previous centuries, as discussed helow.

It is noteworthy that when primitive cultivars or wild species have heen used in modern breeding, in spite of the procedure used being designed to eliminate most of their contributions apart from the speciflc characters for which they were chosen for use, they have frequently contributed useful characters in addition to those intended. This can be viewed as evidence of the limitations of the gene-pool represented by the old varieties, and of the wealth of that which could be drawn upon.

Wild species

About 160 wild species are known, although many of these have not been examined in a living state being known only from herbarium specimens. Many occur in the same areas as the cultivated species although with a wider climatic distribution, from alpine to subtropical; some occur on the plains of Argentina and adjoining countries extending to the Atlantic seaboard, some in Chile, and others in Mexico and neighbouring countries, one or two extending into the south of the United States. Some are closely related to the cultivated potato and others are relatively distantly related, not being directly hyhridizable with modern potatoes.

They have mainly been collected and classifled during the present century. Before that a few had been introduced to Botanic Gardens but taxonomic knowledge was almost non-existant and the few speciesnames which had been applied were used indiscriminately.

For example, Salaman termed a blight-resistant species which he was using in breeding *S. etuberosum* even while acknowledging that it probably was not the species to which Lindley had given that name, and in spite of its bearing tubers contrary to the implication of the name. (It was in fact a natural hybrid between *S. demissum* and *S. tuberosum* which initially arose in the Edinburgh Botanic Garden and was subsequently named *S. x edinense).*

The unreliability of the identiflcations of species used in late 19th-and early 20th-century breeding must be remembered when considering ancestries. Ross has summarized the occurrence of wild species in the ancestries of modern varieties.

These are *S. demissum* and its close relatives, used initially for major-gene resistance to blight but also contributing fleld resistance to that disease and resistance to leaf-roll; *S. vernei* used for eelworm resistance; 5. *spegazzinii* also used for eelworn resistance but contributing resistance to scab and to *Fusarium coeruleum* in addition; S. *stoloniferum* for resistance to *Y* and related viruses; and *S. acaule* used for resistance to virus *X* but also contributing leaf-roll resistance. Breeding from wild species takes time, and various other species are currently in use.

The species are so numerous that it may be dangerous to generalize. However, many have bitter tubers. The tubers are usually small and are borne on long stolons, probably deep in the ground; when wild species are grown in pots they usually forna tubers in the bottom while primitive cultivars form them throughout the pot and often at the surface. The tubers are usually of a round or oval fairly regular shape and anthocyanin pigmentation is rare and is restricted to a faint blueish-purple flush when it occurs.

The range of tuber shapes and of skin and flesh pigmentation found in the primitive cultivars is not reflected in the wild species and therefore would appear to have arisen due to preferential retention of mutations by man. Ross says 'There is no unfavourable trait known which would prohibit the utilization of a wild species hitherto envisaged for introduction into potato breeding.

Even the very high glycoalkaloid content often found in the tubers of wild species could be reduced to normal concentrations after some backcrosses.' But long-term work is clearly involved, and Hawkes calls for 'link men' to undertake the preliminary breeding work and provide good breeding lines, not far from varietal status, to the breeders.

EARLY INTRODUCTIONS

The clearest information concerning early introductions is from descriptions and illustrations by two botanists. In England, Gerard mentioned the potato in 1596 and in the frontispiece of his 1597 Herbal he is seen holding a flowering shoot of the plant. He had apparently been growing it in his garden for some years. Clusius had obtained a different variety in 1588 from Belgium whence it had been obtained from Italy where, he says, it was extensively cultivated.

Gerard said that his potato was from Virginia but this was clearly an error as potatoes did not occur anywhere near there. A possible explanation is that when Drake in 1586 brought back Raleigh's settlers from Virginia, including the botanist Hariot, he may have been carrying potatoes from Cartagena where he had been earlier on the same voyage. Salaman says that by 1549 there was a regular trade route between Bogota in the Colombial highlands and Cartagena by which potatoes could have been brought down, and that while Havana was the main provisioning port for ships bound for Spain, all ships homeward bound from the Isthmus of Panama called first at Cartagena where potatoes could have been held ready for them. Potatoes certainly were used as ships stores; Cavendish found potatoes crated for shipment in storehouses in Chile in 1587.

Drake also saw potatoes on an island off Chile in 1577 and he may have recognized their value as stores but as he was on a round the world voyage he could not have brought them back to Britain in a living state. Thus the most probable source of Gerard's potato is Colombia, brought back by Drake in 1586, though Hawkes questions this and suggests a slightly later date, perhaps 1590, but still with a Colombian/Carribean origin. The origin of the potato described by Clusius is more speculative. It is tempting to assume that it would also be from Colombia, whence introduction to Spain would be easy. However, there is evidence that in 1565 potatoes were sent from Cuzco to Spain at the command of Phillip II and there is also evidence that potatoes were sent by the King of Spain to the Pope at an early date, perhaps 1565. Thus the potato described by Clusius could have been of either Colombian or Peruvian origin. The possibility of a Chilean origin for either of these introductions seems remote. Salaman points out that the harvest there would be in April and the minimum voyage time up the Paciflc coast, by mule and porter across the Isthmus of Panama, then to Spain would be 6 months; planting could not be until the following Spring, a year from harvest with much of that year in tropical conditions.

Survival in a plantable state would be improbable. Apart from that, Clusius' potato had been in cultivation for some time prior to 1588 and the Spaniards had not effectively conquered the Indians in the Chilean potato growing areas until after the probable date of this introduction. There is evidence that potatoes were purchased for use in a hospital in Seville as early as 1573, suggesting an introduction not later than 1570 allowing time for multiplication but Burton suggests that this could have heen the 'sweet potato'. Gerard's potato would almost certainly have been introduced on and English ship, thus being acquired by raiding of a Carribean port or a ship homeward bound from there and the seizure of stores. (One may note that voyages round Cape Horn to Europe were rare; Salaman says the flrst was in 1579 and the ship was cleared of stores before it left the Magellan straits). To assume that the potatoes described by Gerard and Clusius were the only ones grown in Britain and Europe at that time may not be entirely justifled.

Once potatoes commenced being used as ships stores the possibility of casual introductions would exist; these would almost certainly be Colombian/ Venezuelan in origin. Salaman provides descriptions of the two known early introductions based on the original accounts, diagrams and pictures. Both were distinct from modern varieties in' the very irregular shape of the tubers disflgured by outgrowths, and the extremely deep eyes'. They differed in various ways, including among them that Clusius' potato was dark red and very late maturing while Gerard's was white skinned and less late. Salaman argues that all the variation of modern potatoes, prior to the addition of characters in modern breeding, could have been obtained by recombination and selection of genes borne by these two introductions but one feels that the arguments are rather forced and that the variation can be more readily explained if the probability of further introductions is acknowledged.

INTRODUCTIONS AND BREEDING TO THE EARLY 19TH CENTURY

Salaman and Hawkes examined 18 herbarium specimens of British and western European potatoes from between 1600 and 1750. They did not classify them as alike or distinct, and perhaps all 18 were distinct. Perhaps all but three resembled Andean potatoes and differed from modern potatoes in having relatively highly dissected leaves, those with less dissection being from the later part of the period. They attributed the variation to the raising of seedlings, and the tendency towards the modern leaf form in the later ones to selection but, while they quote good evidence for some raising of seedlings in the period, there is no positive evidence that all the specimens were derived from the two introductions mentioned above. Historical evidence for introductions in this period is at best sparse and indefinite though it does appear that John Howard of Cardington, Bedfordshire, 'famous as prison philanthropist, traveller, scientific observer, and a man of determined accuracy', introduced a potato 'fresh from America' in or about 1765 which came to be widely cultivated, remaining in use at least as late as 1840 and acquiring many names including the Howard, Cluster, Conglomerated, Suffolk and Surinam.

But apart from reliance on such sparse items of information one may argue that, as at least two introductions occurred in the 16th century, and as potatoes were being used as ships stores, further casual introductions of Colombian origin must have occurred in the 17th and 18th centuries and, as interest in the potato as food increased, occasional introductions from elsewhere in South America may have been made by enthusiasts. There is no doubt that the number of varieties grown in the early years was limited. Davidson says that the earliest references to varieties were by Rye and Miller; Miller, probably writing in an English context, referred only to red and to white potatoes (which Davidson comments might have been types rather than varieties) but Rye, an Irish writer, said: 'There are five sorts of potatoes known to us: the flat white kidney potato, the round white, the yellow, the round red and the black potato.

The white potato is set in January to afford early potatoes in the latter end of June. The yellow potato is valuable for keeping most part of the summer following. The round red is a good potato and increases much. But it is the black potato (not that the pulp is black, but that the skin is very dark) that is most valued by those who know it; the pulp affords a stronger invigorating diet to the labourer; it keeps till potatoes come again... Since the people of this country found the peculiar goodness of this potato they will scarce cultivate any other.

They will grow so large, as that some of them have measured four inches in diameter.' Davidson indicates that Rye did not comment on the lound white, so the 'early' must have been the flat white kidney. He suggests that the round white and round red may have been types rather than varieties, but that the other three appeared to be distinct varieties. (While Rye was clearly knowledgeable about potatoes, one may assume that he was writing of the Irish situation and one could be mistaken in inferring that only these five types occurred in Europe at that time. However, there seem to have been fewer varieties in England than in Ireland.) The position changed later in the 18th century; the raising of seedlings and selection of varieties became a widespread practice.

According to Salaman, Marshall complained of the indiscriminate raising of seedlings and described the varieties extant as 'endless'. Cartwright took a different view, alluding to the raising of seedlings as 'a valuable corrective to the constant deterioration of existing varieties' although also refraining from listing varieties because of their number, and because such a list could have only a 'local interest'. Two major developments had occurred. Firstly, the potato had become established as a major food crop in England as well as Ireland, and secondly, diseases had become important. Knight wrote that some 66 years previously the potato was 'unknown to the peasantry of Herefordshire'.

It was then grown in small quantities in the gardens of the gentry but was ' not thought to provide wholesome nutriment and was supposed by many to possess deleterious properties'. But within 10 years all prejudices had disappeared and potatoes had almost wholly displaced cabbages from the cottagers' gardens. The diet of the common people had previously consisted of bread, cheese and the produce of their gardens but subsequently the potato was increasingly dominant. He believed that a considerable improvement in general health had resulted; agues previously prevalent, had disappeared, and people were living longer.

At an earlier date he wrote that an acre of potatoes could provide as much food as 8 acres of wheat or 40 of pasture, and expressed alarm at the manner in which the population was expanding on the basis of this crop, especially in Ireland. From other references it appears that extensive cultivation of the potato in England commenced flrst in Lancashire, somewhat earlier than Knight indicates for Herefordshire. A severe condition termed 'the Curl' was flrst recorded in Lancashire in the 1760s and, in the 1770s, it was so troublesome as to threaten potato cultivation there. It was also serious in Yorkshire and elsewhere; in 1770 it caused such loss in Ireland that a large quantity of grain was imported for food.

Knight mentions its having 'a few years ago destroyed many of our best varieties'. It appears from my reading of typed and handwritten notes on old publications assembled by G. H. Pethybridge and preserved in the library of the Scottish Crop Research Institute that the term ' curl' was applied indiscriminately to various serious conditions; infection with potato leaf-roll virus apparently was the most usual cause but frost damage to and perhaps fungal infection of the tubers or ' sets' used for planting and perhaps also planting on unclean land were seemingly implicated in some cases. As Salaman points out, some of the early descriptions seem to refer to mosaic-type viruses rather than to leaf-roll. Nevertheless, severe virus infections were clearly the main cause, and it was soon realized that these were less troublesome in some regions than in others. A trade in seed potatoes developed remarkably rapidly. From notes by Pethybridge on a report of a committee of the Board of Agriculture, 1795: ' In the North Riding of York they procure their seed from the Moorlands, or from Scotland, to avoid the curl; the flrst year free; the second, some curled; the third, almost all.' West Lothian was seemingly free of the disease above 400 ft altitude, but stocks brought down from there became curled after 2 years.

And in an extract from 'Communications to the Board of Agriculture on subjects relative to the Husbandry and Internal Improvement of the Country', 1802, it is mentioned that 'sets' from which curled plants grew did not rot (strongly implicating leaf-roll; similar comments are frequent in the old papers), and 'In this country it has been a practise amongst cultivators... to get their seed from the south-western part of Scotland, which they flnd to be very little, if at all, subject to curl; but the same seed, planted here, will in a year or two be subject to the disorder. In the year 1798 I planted 6 acres with potatoes of this Scotch seed; there was scarcely a curled plant amongst them.

The following year, about an acre and a half was planted with the produce of these very potatoes, and at least one-sixth of the whole was curled.' There is no indication in the notes of the identity of the writer or his region of the country, but the indication that a suflicient quantity had been purchased by a single individual for the planting of 6 acres is noteworthy; moving such a quantity by horse-drawn transport over a long distance would be a major undertaking, but perhaps coastal shipping was involved. But apart from the ' curl' there was a condition - not then recognized as caused by disease — viewed as degeneration or ageing. This seemed to affect all varieties. Knight said that all gradually lost some of their qualities, and quoted the opinion of a Dr Hunter that their duration ' in a state of perfection' averaged about 14 years; in 1833 he indicated that they always declined in yield and sometimes in quality also.

Consequently, varieties were viewed as of only transient value and in 1814 Knight said ' New varieties of the potato are so readily obtained from seeds that the preservation of any one now cultivated is an object deserving little attention.' Knight stressed that to obtain high yields it was necessary to use a new variety; Shirreff (before 1819, from notes by Pethybridge) said ' To obtain vigorous plants and secure productive crops, recourse must frequently be had to varieties newly raised from seed.' According to McKay the first mention of degeneration was by Maxwell in 1757, but McKay did not clearly differentiate between the 'curl' and the slower and more insidious form of degeneration which the old writers viewed as senility, and it is uncertain to which Maxwell referred. There is no indication in the old papers I have seen of any association in the writers' minds between the ageing of a variety and the 'curl', nor any mention of symptoms of any nature being associated with the ageing process. Nevertheless, one may assume that such degeneration arose from infection with some disease or diseases, transmissible by tubers, of a milder nature than those causing the severe symptoms of ' curl'.

The proliferation of varieties in the late 18th and early 19th centuries apparently arose from the raising of seedlings, not from new introductions. In 1810 Knight said that most varieties grown in Herefordshire were of Irish origin and were ill-adapted to local conditions. He pointed out that in Ireland they were secure from frost from the end of April to the near end of November but the safe period in England was much shorter, and the Irish varieties, grown in his area, were almost always killed while in full bloom. There was some exchange of varieties between countries; Knight in the 1830s had one from Flanders and the Horticultural Society of London had some with French names in their garden.

But there is no evidence that new South American introductions were contributing to the range of material in cultivation. Seedlings can be raised in quantity and there would be ample opportunity for selection among them; a general raising of standards as regards day-length adaptation and tuber sizes and shapes would result. While selection would usually be incidental to the raising of seedlings to counter ' degeneration' there is one indication of the raising of seedlings specifically for selection. Hawkes quotes from Miller who described selection, by cultivators in the Manchester area, for earlier tubering.

Berries from the earliestfiowering plants were used for sowing and by frequently repeating this process varieties were obtained 'fit for use in two months after planting'. The ' earlies' grown in the early 19th century were not comparable with modern 'earlies'. Knight described them as dwarfish plants, unsuited to extensive field culture, grown in hot beds frames or garden beds at very close spacings, giving small yields of very small tubers but being only a short time in the ground.

They were delicacies for the tables of the gentry and are to be seen in the context of the lengths they seemingly were prepared to go to to obtain 'young' potatoes out of season, for example, burrying stored potatoes in sand to induce formation of new tubers directly. It appears from Knight's writings that no intermediates occurred between these and varieties which, in English conditions, failed to mature naturally. Hawkes indicates that' first earlies' as presently known did not appear until the second-half of the 19th century and suggests that they all derive from the variety Early Rose which in turn derives from Rough Purple Chili, introduced in the 1850s. (Where Rye's 'early' fits into the picture is vinclear, but one may note that it was accorded a 6-month growing period in the milder climate of Ireland. One may also note that ' earlies' appear to have been such as would give an early crop, and they may not have died off early). There is some difficulty in achieving a balance between the picture of massive proliferation of varieties of mainly local cultivation and of transient interest and a contrasting picture of a few outstanding varieties grown widely for long periods of time.

Davidson indicates that the yellow potato mentioned by Rye in 1730 was later mentioned by Hale (published in 1756 but written in 1746) then again by Lawson in 1836, while the black potato was recommended by Howden as late as 1837 and over the period 1795 to 1824 was mentioned in 11 County Statistical Surveys, six Irish and the others Scottish, northern English, and the Isle of Man. Various subsequent varieties attained wide and lasting popularity. There is a distinct possibility that resistance to major viruses was a factor in several cases • Davidson comments 'It is significant that in reference to these outstanding varieties the remark that they "never curl" is so often made.'

However, whether they suffered the general degeneration which then was viewed as 'ageing' is not clear; from the various comments that all varieties degenerated it seems likely that they did so, like the rest. In these early years potatoes were often grown for stockfeed. Clusius said that the Italians used the potato both for human consumption and as food for pigs, and among the historically important varieties discussed by Davidson, while most were grown for human consumption, Howard's potato, the Ox Noble, the Yam and the Lumper were grown mainly for stockfeed, being high yielders but of poor quality, some being ill-fiavoured. Knight urged that a part of the crop should be fed to stock in order that meat could be included in the peasant's diet.

Yields obtained in commercial practice ranged from 4 to 12 t acre~ ^, 'never exceeding 12', and claims by Knight of yields in the order of 18 to 241, using his system of culture, led to experiments in the garden of the Horticultural Society of London reported on by Lindley from which the main conclusion was that it was 'quite practicable to at least double the crops that are usually obtained'. Knight advised planting relatively deeply, using uncut and relatively large, sound tubers spaced sufficiently widely to avoid over-crowding of the foliage, and planting relatively early to secure as long a growing period as possible; in 1833, in reporting a yield of over 34 t, he stressed that a new variety was required. Lindley referred to ' - methods at present in use, of planting sets or fragments or the smallest tubers or even as sometimes in Ireland, parings of the potato - '. But the use of sets was widely considered to be superior to use of whole tubers and subsequent experiments appeared to confirm this. Cutting the seed, though often done abroad, is so foreign to modern British practise that one may be surprised to find that it was ever done here; however, Salaman says of the variety Majestic ' It is always recommended that this variety be planted whole for seed purposes' which suggests that the practice of cutting was not unknown in Britain as late as the early 20th century.

THE 19TH CENTURY

Potato introductions

During the 19th century there was an increase in the frequency of documented introductions, but this was accompanied by critical appraisal of the material introduced, leading to most of it being discarded. Only two introductions, Daber and Rough Purple Chile, are definitely known to have found favour and to have become the parents of varieties, although there may have been others. In the first-half of the century the Horticultural Society of London was concerned at the situation regarding the potato.

The alarm expressed by their President, Thomas Andrew Knight at the expansion of population that was taking place on the basis of this food supply has been mentioned above. Their Secretary, Joseph Sabine, in 1822 developed Knight's earlier comments, saying that, with the exception of wheat and rice, the potato had become 'the vegetable most employed as the food of Man' with a prospect of soon becoming the chief staple of life, and that this could lead to a great extension of population, though not necessarily to an increase in the happiness of mankind especially as the potato was liable to ' injury from casualties of season' and it was not known how to store it beyond a few months. He warned that' a general failure of a year's crop, whenever it shall have become the chief or sole support of a country, must inevitably lead to all the misery of famine, more dreadful in proportion to the numbers exposed to its ravages'. He urged that the Society should attend to selecting and obtaining varieties ' not only with superior qualities in flavour and productiveness, but which shall be less subject to injury by changes of weather when in growth, and which may possess the quality of keeping for a length of time...'.As one aspect of this, he urged the seeking of further introductions from South America.

In 1831 Knight again mentioned the expansion of population, commenting to the effect that while this was a problem it would be a greater calamity if that population existed without the potato to support it. However, while he clearly remained nervous, he thought the situation to be relatively secure due to the large number of varieties grown, which seemed unlikely all to fail in the same year. Sabine's reference to ' injury by changes of weather' is intriguing and may refer to disease epidemics, but this was long before the blight epidemics of the 1840s and probably before the appearance of that disease in Britain. Lindley indicated in 1848 that Chilean potatoes had not been brought to Europe since the potato disease broke out so it would seem that blight was not known in the 1820s when Chileans were imported. From Sabine, Holford and Lindley it appears that two Chilean varieties were obtained in 1822 and a few clones of the Golden Potato of Peru in 1823, 1824 and 1846, Cultivated potatoes were obtained at various dates from Mexico (said to have originated from Peru), from New Grenada and from the Azores. By 1826 a number of unusual types obtained from British or European sources were under study (the Asparagus potato, the Mouse potato, the Pine Apple or Cone potato and the Spanish Dwarf potato); from their descriptions it seems probable that they were primitive cultivars of South American origin.

A few wild species had also been obtained. None of these were recommended for commercial use. Ross gives 1830 as the date of introduction of Daber to Europe and says it was round-oval, white fleshed, wart immune and had a high starch content. Hawkes suggested that it was probably Chilean in origin. Goodrich introduced 12 clones to the USA over the period 1848 to 1852, although two of them appeared identical. All but one were too late maturing, most by probably 2 or 3 months, and were eventually discarded together with their progenies. The exception was that which he named Rough Purple Chili. It was one of eight obtained from Panama in 1851 and 'supposed to have been brought from the coast of Chile'. Goodrich became somewhat disiuUusioned, and finally questioned ' the propriety of importing sorts fromi climates very different fronn our own'. In Scotland Paterson, commencing in the 1850s, introduced potatoes from many countries and while it appears that to a large extent he was re-assembling material previously distributed from Britain and Europe, Mclntosh says that his sources included Central America and Chile. There seems to be no means of knowing whether any such true introductions contributed to the varieties released by him. Sutton grew the

Golden Potato of Peru for eight seasons but concluded that it was of no use under British conditions due to low yield and disease susceptibility. It would appear that the general raising of standards of performance of potatoes which occurred in the second-half of the 18th century had been such that new introductions from South America generally appeared primitive.

Only rare exceptions, of which we can positively identify only Daber and Rough Purple Chili, performed well enough to be retained and used as parents. Of these. Rough Purple Chili was said by Hougas to have given rise to over 250 varieties; this, by now, is probably an underestimate because many modern varieties and breeding lines have it in their ancestries. For example, through Abundance, Epicure and Pepo it entered the Scottish Plant Breeding Station programme at the earliest stages and is in the ancestries of almost all varieties emanating from there. Daber is said to have given rise to almost as many varieties as Rough Purple Chili. It was used mainly on the Continent of Europe but as European varieties came to be used in breeding in the USA it acquired an infiuence there also and, through the Polish variety Busola which is in the ancestry of the US variety Katahdin, it enters the ancestry of some British varieties, e.g. Pentland Crown and Croft.

Effects of the blight epidemics

It is not my purpose to discuss the social effects of the blight epidemics but one may note that in Ireland, where dependance on the potato was very great, of a population of about 8 million, about 1 million died and 1-5 million emigrated between 1845 and 1860. Sabine's and Knight's warnings of the dangers of dependance on the potato were shown, unfortunately, to have been only too well justified. The precise date and manner of introduction of blight to Europe is not known but as it is a Mexican disease it probably came with potatoes from that country.

Davidson says that almost all the old outstanding varieties, except the Regent, were swept out of existence by the 1846 epidemic. He specifically mentions the Black Potato and the Irish Apple and says that it was the failure of the Lumper, the greatest cropper of its time, that primarily caused the famine. Though viewed by farmers as mainly a stockfeed potato, richer people preferring the higher quality of the Cups variety, it appears that the Lumper was widely grown by the poor for domestic consumption. An upsurge in potato breeding followed. Paterson had as his aim the improvement of blight resistance. Mclntosh says that Goodrich thought that disease susceptibility arose from potatoes having been weakened by long continued asexual propagation and they could be rejuvenated only through sexual reproduction. Superflcially this seems at variance with Goodrich's practice, the introduction of exotic material followed by raising from naturally-set seed, and one could think that his concept was rather that exotic material might prove more resistant; however, it could well be that the varieties available in the United States at that time were sterile. Lindley was looking for resistance in the exotic material available to him, with discouraging results.

These mid-century introducers and breeders perhaps had altruistic motives but later in the century breeding became very much a commercial activity. The degree of emphasis then placed in blight resistance is not clear; in Findlay's descriptions of his varieties, disease resistance is sometimes stressed but more frequently is mentioned, if at all, with no greater emphasis than yield, appearance, maturity and table quality. However, high susceptibility would clearly rule out a variety and one may assume that varieties available for use as parents were not especially susceptible, all highly susceptible varieties having been eliminated by the epidemics. It should not be thought that the blight epidemics were a feature of the middle of the century only. Salaman says of Magnum Bonum and Champion that they ' did a great deal to ameliorate the damage which the blight again produced in 1879, and to which every other variety fell a hopeless victim'.

Breeding methods

Thomas Andrew Knight may have been the flrst potato breeder, perhaps one of the flrst breeders of any crop, in anything approaching a modern sense. In 1807 he proposed cross-pollination with a view to seeking to combine characters, and in 1810 and again in 1831 he claimed to have succeeded in this. In 1807 he said that earlies never flowered - he suspected due to the ' preternaturally early formation of tubers'; they were obtained accidentally among seedlings raised from later kinds. He described a method of forcing them into flower, consisting of tying them to stakes then washing away the soil from the base of the stem so that tuber formation could be prevented.

He wrote ' I t appears not improbable that by introducing the farina of the small and very early varieties into the blossoms of those of larger size and somewhat later habits moderately early varieties, adapted to'fleld culture and winter use, might be obtained, and the value of these to the farmers in the colder parts of the Kingdom whose crops of potatoes are succeeded by one of winter wheat would be very great.' In 1810, he said that his conjecture of 1807 that varieties could be obtained of luxuriant growth and adapted to fleld culture, ready for lifting by the end of August or early September allowing ample time to prepare the land for winter wheat, was conflrmed. He also believed that de-blossoming increased yield ' particularly if the variety be one which produces seeds', and that non-flowering varieties might be superior. In 1831 he said that by cross breeding between earlies 'of dwarflsh growth and therefore improper for extensive fleld culture' and varieties of tall luxuriant growth ' I can communicate to the latter the habit of producing tubers only, without blossom.' He said he then had a good many such varieties, though selected from a very great number which were totally worthless, and that many which did not produce blossom had other defects; some were weak stemmed and were beaten down by rain and wind; some had low specific gravities and were 'aqueous and worthless'; and some had red veins in the flesh. But he had several which did not blossom and were without obvious defects, though he w^s ' far from thinking that I possess any... which approximate to the greatest state of perfection which the species is capable of attaining'. He had some which Commenced sprouting early in the spring and others later; the early ' vegetaters' were generally higher yielding due to their longer growth period, but the later ones were useful in remaining good until late in the spring when good vegetables were scarce. He also had some which vegetated late in spring yet matured fairly early in the autumn.

He seems not to have given his varieties names; when grown in the Society's garden they were referred to by numbers. And he seems to have distributed them free of charge in small quantities for multiplication by the grower; in 1817 he sent three for distribution through the Society which he said supplied his table in succession through the year, one used soon after harvest, a second for winter and spring consumption, and a third which ' remains perfect longer in the spring and summer than any other I have met with'. Although Knight was cross-pollinating before 1810 this does not seem to have become general practice until the second half of the 19th century, and the raising of seed from naturally-set berries continued, alongside deliberate crossing, into the early 20th century. The last successful variety obtained by natural pollination and known to me was McKelvie's Arran Victory, raised in 1912 and issued in 1918.

William Paterson of Dundee, who commenced breeding in 1853, relied on natural pollination but in plantings containing many varieties intermingled with a particular one, the Rock potato, presumably in the hope of obtaining crosses. While I know of no other deliberate attempt to use natural pollination in this way, one may assume that individuals especially interested in varieties and selection of new ones would grow many, the possibility of outcrossing thus existing. However, some comments on the probability of crossing may be useful. Findlay thought that naturally-set berries derive entirely from unaided selfing, saying that he had only occasionally seen ' a humble bee alight on the edge of a petal, but never saw one explore the bloom'. This seems broadly true of most modern varieties which have little pollen, but our experience with highly fertile potatoes such as Neo-Tuberosum is very different; bumble bees eagerly visit them to collect pollen. But even so, the pollen deposited on the style as a result of a bumble bee visit is usually mainly the flower's own, with some from other flowers in addition, the resultant seed being a mixture of selfs and crosses. In one experiment about 80 % of the seed had been set as a result of selfing. It is common experience that potato varieties which lack fertile pollen may yet be female fertile. Such varieties would have no attractions for bees, but in mixed plantings containing pollen-productive varieties an occasional unproductive flower may be visited, in exploration or in error, and seed may be set; such seed would necessarily be resultant on cross-pollination. Crosses tend to be more vigorous than selfs, and in mixed seed there may be a tendency for crosses to be selected preferentially. But, in as far as pollen fertility reflects the outbred status of the parent, selfed seedlings from a highly fertile parent may be little inferior in vigour to crossed seedlings obtained from it and be quite likely to be selected, while among the seedlings of a relatively inbred parent those derived by crossing may be much superior to those derived by selfing.

Thus, broadly speaking, among selections made from naturally-set seed from mixed plantings including varieties which are pollen-fertile and attractive to bumble bees, those derived from the more fertile mother-plants are likely to be selfed derivatives while those from less fertile parents are likely to be the result of outcrosses. But in single-variety plantings, or in mixed plantings of relatively sterile varieties providing too little pollen to attract the bees, inbreeding will be the rule. The early potato introductions were fertile, berries being shown in the illustrations, and fertility must have been high in the 18th century varieties to permit the widespread raising of seedlings and selection of new varieties which commenced then. But it appears that by the mid 19th century the population was relatively sterile. McIntosh says that Paterson sowed seed from those relatively few of the varieties in his mixed plantings which bore berries, and he also says of the Fluke, which is believed to have been the mother-plant of Paterson's Victoria, that it was ' about the only variety grown at that time which bore natural berries'. There is thus a question as to whether Paterson's plantings, and those of subsequent breeders, bore sufficient pollen to be attractive to bumble bees or whether such berries as were obtained from them derived from unaided selfing.

But one is inclined to assume that Paterson, in choosing a variety with which to intermingle the others, would have considered its fertility. There is a question as to why fertility had declined. One reason could be the deliberate selection of non-berrying varieties, as urged and practised by Knight, but another reason could well have been inbreeding. With large numbers of farmers ' rejuvenating' their stocks by raising seedlings, taking seed from their own fields which usually would contain single or relatively few varieties, there must have been considerable inbreeding. Several, perhaps most, breeders subsequent to Paterson and Goodrich operated by crossing. Some did their own crossing while others selected from seed supplied to them; for example (though this was in the early 20th century) McKelvie selected Arran Chief, Arran Comrade and Arran Rose from seed supplied by another breeder. In some cases crossing was random; for example Miles selected Great Scot from seed obtained by 'fertilizing with the mixed pollen of several plants the stigma of one or two female parents' while the batch of seed from which Arran Comrade and Arran Rose derived was supplied with the information ' 100 crosses in this, about 30 table varieties and all the best 1910 varieties'. Many breeders either did not maintain records, or lost them, or regarded them as 'trade secrets' not to be disclosed. Thus the ancestries of old varieties are frequently unknown.

And unfortunately, when they are supposedly known they cannot always be relied upon. Thus Findlay, who did not normally reveal parentage, did so for Up-to-Date, saying according to Salaman that it was from Paterson's Victoria x Blue Don although in his own book he says that 'one of the parents was never sent out by me or given by me to anyone' and its parentage is given in Stegemann and Loeschcke as Paterson's Victoria crossed with a self-set of Blue Don; there is though some question as to whether Findlay did breed Up-to-Date because, although it was a very distinctive variety, it appeared indistinguishable from the variety Scottish Triumph bred by Gemmel which was exhibited 2 years before Up-to-Date was released.There are doubts as regards Findlay being the true originator of some of the other varieties issued by him; thus the Bruce, for which he also gave a pedigree as being from Paterson's Victoria x Late Rose, appeared identical with Magnum Bonum, his Jeannie Deans appeared to be Abundance, and he re-issued one of his own varieties, Evergood, under the name Eldorado (a financial gamble centred on this and another of Findlay's varieties, £100 being paid for a single tuber of Eldorado, leading to ruin).

My purpose in not to single out Findlay for criticism but to point out the danger of assuming stated pedigrees to be correct. Many people were in the business of re-naming varieties and passing them off as being distinct (though perhaps not of claiming to have bred them and attributing pedigrees to them). Varieties obtained many synonyms; the extreme example seems to have been Up-to-Date for which Salaman lists 186 alternative names. Other problems regarding pedigrees arise from names being re-used. Thus Findlay named one of his varieties Great Scot many years before Ezra Miles bred the variety of that name mentioned above; Paterson issued a British Queen as also, many years later, did Findlay; and Suttons Ltd., issued two varieties with the name Flourball, one in 1870 and the other in 1895. Also, Nichol of Arbroath bred the variety Champion, issued 1879, which became important as a blight resister, while a variety of the same name was grown by the Horticultural Society of London in the 1830s.

In these cases one may perhaps assume that the earlier variety had gone out of cultivation before the name was re-used. Some breeders seem to have had favourite parents which they used extensively. Thus Findlay said 'Individually, for breeding purposes, I would not give a farthing for any potato if I could not trace its descent from the Victoria, either on the male or the female side.' And according to Stegemann and Loeschcke the parentage of his

British Queen was the same as that which he claimed for Up-to-Date, though Salaman says the parentage of British Queen was not disclosed. Regarding Paterson's Victoria, Salaman says ' Indeed, it may be said that practically no potato today of any outstanding merit is without the blood - though far removed — of this variety.' It seems probable, therefore, with many varieties having been eliminated by blight and most being fairly sterile, and with breeders having favourite parents, that relatively few varieties contributed as parents to subsequent generations of varieties. Breeders seem generally to have used named varieties as parents, breeding lines being almost unknown.

With the exceptions of the imported variety Adirondack and his own variety Barley Bounty, all parentages recorded iri Salamari's book involve only named varieties. But Findlay may have been an exception. While there are doubts about some of his varieties, there seems no doubt that he was genuinely a breeder and his indications as to his breeding methods may have substance. He said ' I keep my breeding stock... I sell their progeny, but never sell from this stock' and ' I think the art of selection will ever be the occupation of a very few. I do not refer to the selection of the potatoes after you have grown them from seed; I mean the working up of a reliable breeding stock.'

Some comments on the breeding of the variety Majestic, released by Findlay in 1911 and one of the last of his varieties, seem desirable because it dominated the British maincrop acreage for many years, it is in the pedigrees of subsequent varieties, and intriguing questions regarding it have arisen. Salaman says its parentage was not disclosed, in common with others of the 13 bred by Findlay for which he gives descriptions apart from Up-to-Date (and the Bruce which he does not describe but mentions in a footnote).

However, Stegemann and Loeschcke say it was from an unknown female parent crossed with British Queen and Ross goes further, indicating that the unknown parent was a 'wild potato' and suggesting that it might have been *S. xedinense;* in personal correspondence he says that the indication tbat it was a wild potato comes from Siebeneick but he does not know how Siebeneick obtained the information. The great and long-lasting success of this variety suggests that it may not have been entirely ordinary, and that the indication that there was something unusual in its breeding is not to be lightly dismissed. However, Findlay was not a scientist and one hesitates to think that he would introduce anything truly wild to his breeding.

There is no other indication that he did so, and in bis 1905 book he appears to dismiss the possibility of using wild species although his attitude to primitive cultivars, which he regarded as potatoes 'in a wild state', appears more positive. There does seem, therefore, to be a possibility that Majestic may have a primitive cultivar in its parentage or near ancestry. Majestic does not have i?-genes for resistance to blight which would be expected to be present if it had been bred from *S.* x *edinense.*

THE SITUATION IN THE EARLY 20TH CENTURY

Bukasov commented that potato breeding had been stewing in its own juice, always using the same parents in innumerable combinations, leaving many problems such as blight and virus diseases unsolved; that only isolated random specimens had been introduced and interbreeding had led to an unbroken ring of nearly allied crosses, and a cul-de-sac. Eighteenth and early 19th century breeding, in response to virus epidemics, while leading to improved adaptation of the potato to the British climate and requirements, probably reduced the genetic base of the crop and also, because of the degree of inbreeding involved, led to relative sterility. It may well be that the significance of Rough Purple Chili, when it was introduced in the mid-19th century, lay at least in part in its fertility and in the opportunity it offered for a resumption of active breeding, breeding that was urgently needed to replace the varieties eliminated by the blight epidemics.

Those epidemics had further reduced the genetic base, and one may note Davidson's comment that all but one of the old outstanding varieties, which were probably virus resisters, were eliminated. Breeding later in the 19th century although involving some exchange of parents between countries (especially of parents bred from Rough Purple Chili) was mainly with local stocks and, although it was largely by crossing, seems again to have led to relative sterility as indicated by Findlay's comment on the lack of interest of potato flowers to bumble bees. The situation may have been worst in the United States where Hougas comments that 'no new varieties of consequence other than clonal selections were released during the period 1900 to 1932' and, referring to the pollen fertility of the British and Polish varieties Sutton's Flourball and Busola introduced in the early 20th century, says 'This character, without doubt, is one of the primary factors which determined their initial selection as parents since the old American varieties available to 20th century breeders produced, with few exceptions, no or very limited quantities of functional pollen.'

Nineteenth century breeders, under the influence of Knight, may have preferred non-berrying selections on the assumption that production of berries detracted from production of tubers; however it seems at least equally likely that the loss of fertility was due to inbreeding deriving from close relationships between the parents used. One should also note the opinion of Grun that 19th century Chilean introductions had a cytoplasm which, when combined with the chromosomes of the old varieties, induced pollen sterility. Sterility still places constraints on breeding and, while it is usually possible to induce flowering in a desired parent and use it as female, the choice of suitable male parents is frequently limited. Two useful sources of information on the situation early this century are Salaman's book 'Potato Varieties' for Britain and Clarke and Lombard's booklet 'Descriptions of and key to American potato varieties' for the United States. The appearance of wart disease, at the turn of the century, was followed by the observation that some varieties were immune.

This, together I believe with the need to maximize production of home-grown food in the 1914 to 1918 war, led to the establishment of a committee under Salaman's chairmanship to enquire into the varieties of potato grown in Britain. His book summarizes the resultant knowledge. He lists about 500 distinct varieties and gives histories and descriptions of 89 which were in more or less general cultivation. Many of the 500 had alternative names; as already mentioned, the extreme was the variety Up-to-Date which was found to have 186 other names. One may summarize the origins of the 89 widely-grown varieties as follows. Seven were believed to have arisen as sports; three of these and eight others were imported, five from the USA and six from Europe. The other 74 presumably arose as seedlings in Britain. Of the 74, five were known to be from open pollination, 42 were of unknown breeding and three doubtful, and 24 were from crossing. Of the five from open pollination, the seed-parent was known or suspected in three cases.

Sixteen of those of unknown breeding were by breeders who obtained other varieties by crossing and may have obtained these by that method also. The three ' doubtfuls' consisted of Up-to-Date, commented on above, and Magnum Bonum and Langworthy which were obtained by Clarke from the same berry of Early Rose but there was uncertainty as to whether that berry had been openpollinated or had been crossed with Paterson's Victoria. Of the 24 obtained by crossing only 14 had both parents known, the uncertainties generally arising from the sowing of mixed progenies. In spite of the paucity of information on parentage it is apparent that Rough Purple Chili was in the ancestry of eight British varieties and a sport of one of them, and in three of the importations from the United States. A European variety was in one ancestry, and a New Zealand variety in another. The wild species *S. commersonii* was, according to one account, a parent of one variety, but Salaman preferred another suggested parentage for this.

In all other instances the parents, where known, were British and, as noted above, Salaman almost invariably shows them as being named varieties. The general picture therefore is of breeding by open pollination or by crossing, the parents almost invariably being named varieties and, apart from a significant 'input' from Rough Purple Chili, usually being British. Clark and Lombard list 53 varieties cultivated in the USA at that time and Folsom gives further information on the breeding of many of them. Two were imported. Of the others, the origins of 14 were unknown; presumably these would involve only North American varieties in their ancestries, these in turn tracing to old introductions from Britain and Europe. The remaining 37 were of known or suspected origins and these traced to only 15 parents. The latter consisted of seven old North American varieties and an unknown which would probably be an eighth such, one selected in North America from seed obtained in Europe, two British and two European varieties, one Costa Rican, and Rough Purple Chili. The Costa Rican introduction was ancestral to only one variety but Rough Purple Chili was in the pedigrees of 21 of the 37 of known derivation. No non-cultivated species was involved, and apart from the Costa Rican and the Chilean ' inputs' the ancestries of all the varieties seemed to be rooted entirely in the long-standing potato stocks of Europe and North America. The 37 of known parentage consisted of 11 sports and 12 derived by open pollination of named varieties (one by selfing after open pollination), eight crosses between named varieties, three crosses of named varieties with numbered breeding lines, and three inter-crosses of breeding lines. Those involving breeding lines were relatively recent.

This picture of the early 20th century situation in the United States, which of course is 20 years subsequent to the picture of the British situation derived from Salaman's book, differs in the greater involvement of 'exchanged' parents, in the apparently greater significance of Rough Purple Chili, and in the increasing use of breeding lines. But it is very similar in that it involves varieties obtained both by open pollination and by crossing, crossing still being mainly between named varieties, and in the absence of any involvement of wild species and the virtual absence of primitive cultivar ' inputs' other than Rough Purple Chili. As regards the use of' exchanged' parents one reason, the requirement for pollen fertility, has already been mentioned. Another reason was introduction of disease resistances. Cooking quality and yielding ability are also mentioned in relation to some of them. Varieties involving 'exchanged' parents were issued from 1932 onwards.

A COLLABORATIVE BREEDING STRATEGY FOR ORGANIC POTATOES

Average potato yields in Dutch organic farming systems are rather low and variable compared to Dutch conventional farming standards. The main reason for this year-to-year variation in yield is due to the disease known as late blight, caused by the pathogen *Phytophthora infestans,* the timing and severity of infection is unpredictable and very variable. In the last few years late blight has been so devastating in organic potato production that its acreage has decreased by 20%. The ability of the organic farmer's agronomic toolbox to reduce the susceptibility of a potato crop to late blight is limited, certainly under the Dutch climatic conditions. Conventional breeding programs do not generate either a sufficient number nor appropriate types of resistant varieties for organic systems; and a separate breeding program for the limited area of organic potato production is simply not viable. Therefore, all concerned forces have joined together in an umbrella program called *Bioimpuls* for organic potato improvement. Within this program there is a joint breeding project aimed at developing better adapted and late blight resistant varieties. With financial support from the Dutch Government this organic breeding program started in 2008. It is managed through the cooperation of six potato breeding companies, organic farmers and breeding scientists from the Louis Bolk Institute and Wageningen University.

COLLABORATIVE BREEDING STRATEGY STRUCTURE

The organic breeding program has built on an successful collaborative structure between potato farmers and company breeders that has existed in the Netherlands for more than a hundred years, a set up that is hardly known in other European countries. In this system each farmer-breeder discusses individually with the company breeder which combinations of traits they consider are desirable and which parent lines should be crossed to have the greatest chance of meeting the breeding goals. In this way the farmer's experience in growing potato varieties and the breeder's experience in heritability of traits are integrated.

The company breeder is better equipped to conduct the crossings in the greenhouse and to collect the seeds. The following spring the seedlings are raised by the company breeders and offered to the farmerbreeder to grow on. On average, a farmer-breeder yearly plants 500-1,000 new seedlings, though some can cope with more. Each seed can be the basis of one potential variety. In the first year the seedling is planted and its progeny are then evaluated and selected or discarded. After the first year, the selected tubers are multiplied vegetatively, ensuring that there is no further segregation and therefore the selection is relatively easy for farmers. Farmer-breeders select for basic agronomic characteristics such as foliage and tuber appearance, length of stolons, tuber size and distribution, adding their own insights and preferences. The selection percentage is individual, but can for instance be 15% in the 1st year, 33% in the 2nd year, and 0-30% in the 3rd year. The company breeder will proceed with the clones that are selected by the farmer-breeder and will test them on further quantity and quality traits, and might end up with one potential variety that will be sent in for the official trials for variety registration.

CONTRACT FOR SHARING ROYALTIES

At the start of a collaboration between a breeding company and a farmer-breeder a contract is signed arranging the legal aspects of the collaboration, such as ownership of the genetic resources and sharing of the royalties. If a farmer-breeder's selection is successful and ends up as a marketable variety, it will be registered in the name of the farmer-breeder, but the company will represent the variety in the market and will conduct its maintenance. Therefore, the company will collect the royalties and will keep a yearly account of the received royalties based on the volume of seed potatoes sold. The sharing of the royalties is usually on a 50-50% base, depending on the amount of work conducted by each party. The yearly selection effort of the farmer-breeders has contributed to an increase in the number of potato clones selected each year and therefore the chance of success. Thus, this system has, over the past years, achieved great progress in the Dutch potato breeding sector, with about 50% of the released Dutch varieties being a result of this collaborative system.

Table. Sharing of Breeding activities in a collaborative model of potato breeding in the Netherlands.

Year	Activity	Company Breeder	Farmer-Breeder
0	Choice of parents	×	×
1	Crossing and harvesting seeds	×	
2	Sowing and raising of 1st year seedlings	×	
2-4	Visual selection in the clones for basic agronomic characteristics		×
5-8	Trials for production, resistances, quality and adaptation	×	
5-12	Yearly visit and discussion on progress	×	×
9-12	Research for potential market, registration on nationalj variety list, obtaining plant breeder's rights	×	
13-15	Market introduction, maintenance, collecting of royalties	×	
13-	Yearly accounts and sharing of royalties	×	×

BREEDING FOR ORGANIC AGRICULTURE

Training courses

One of the aims of the *Bioimpuls* organic breeding program is to increase the number of farmerbreeders involved. The more farmer-breeders associated with a breeding strategy based on a 'no cure/no pay' basis, the more economically feasible the breeding program becomes. Before the start of the project only two of the 160 potato farmer-breeders active in the Netherlands were organic farmers.

After 14 years of effort, one of these two farmers was recently successful with the selection of the variety called Bionica, which in 2007 was marketed by the associated breeding company C. Meijer B.V. In spring 2008 and 2009 the first organic breeding courses to introduce the basic principles of potato breeding and selection were organised for interested farmer-breeders. The course gives farmers more insight into the breeding process and helps them realize that such a long-term activity needs commitment, skill and time; and that, therefore, must be given organizational space within their farm management.

Currently, a total of eight organic farmer-breeders have joined the organic breeding program where the farmers will be supervised by an experienced potato breeder.

Pre-breeding

Next to the breeding activities of farmer-breeders, there is also a need for a long-term pre-breeding activity to generate new genetic material, including new sources of late blight resistance by using material that was identified in earlier projects by Wageningen University. Therefore, specific crossings were made, including the introduction of new sources of resistance found in the species *Solanum bulbocastanum, S. berthaultii, and S. okadae.*

It is planned that as much as possible these new sources of resistance should be integrated into a genetical background that includes other required traits for organic potato, such as nitrogen-efficiency, resistance to *Rhizoctonia* and early blight, and early tuber setting and bulking.

Collaboration between central organic breeding program and commercial breeding The organic potato breeding program now includes the cooperation of six commercial breeding companies that are already involved in the propagation of organic seed potatoes. The farmer- breeders are all attached to one of these companies.

Each year the farmer-breeders receive seedlings from the central organic breeding program, conducted by Louis Bolk Institute and Wageningen University, and/or from their commercial breeding company. The breeding companies can produce crosses with their own selected parents as well as derive new genetical sources from the central organic breeding program.

Through a contract between the central breeding program and the involved farmer-breeders and breeding companies, it is agreed that in future when varieties derived from plant material from the central breeding program are registered, 10% of the royalties will flow back to the program. This means that the program will be able to partly generate its own income for future breeding activities.

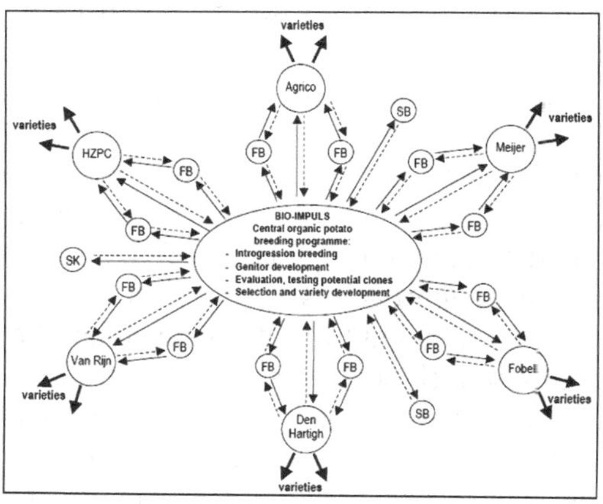

Fig. Overview of the collaboration between the central organic potato breeding program Bioimpul, the farmer-breeders (FB), some independent small breeders (SB), and named commercial breeding companies in the Netherlands, 2009.

FUTURE

An organic breeding program as described in this article can only be successful and economically feasible when the number of organic farmer-breeders increases. It is estimated that at least 20,000 seedlings per year need to be tested for the program to be viable. Potato breeding is a longterm activity and is a continuous search for new resistance sources.

The recent release of a new late blight resistant variety shows that with the efforts of more organic farmer-breeders the selection of a range of varieties suitable for organic potato production in the future is achievable. However, the resistance traits that will be used in the organic breeding program are based on racespecific *R*-genes.

This kind of resistance is assumed to be not durable. Therefore, the project's program will include pyramiding of two to three genes from different sources in many different combinations. However, in the field one can only see whether a clone is resistant or not, and not whether the resistance is caused by one, two or three genes. In the future, marker assisted breeding using freely accessible markers, will be a necessary instrument to achieve such adequate pyramiding.

CLASSICAL GENETICS AND TRADITIONAL BREEDING

GENETIC CHARACTERISTICS

The cultivated commercial potato *Solanum tuberosum* is a polysomic tetraploid crop with basic chromosome number $x = 12$, and a haploid genome size of approximately 840 Mbp, which makes it a mediumsized plant genome. Each of the four copies of homologous chromosomes segregates from multivalents or random bivalents. Due to the small size (about 1–3.5 μm at mitotic metaphase) of individual chromosomes, it has been generally regarded as a cytologically difficult species to study.

Indeed, potato mitotic chromosomes cannot be identified based on their morphology. Mohanty et al. found high variability in cytological characteristics (e.g., total chromosome length and volume, number of secondary restricted chromosomes) between various cultivars and breeding lines. On the basis of size and position of constrictions, they classified potato mitotic chromosomes into four groups. Advanced approaches are now the most suitable alternatives to the strategies traditionally used for potato chromosome analysis.

Reliable identification of the potato somatic metaphase chromosomes has been achieved using a set of restriction fragment length polymorphism marker-anchored bacterial artificial chromosome (BAC) clones as florescence in situ hybridization probes. This work enabled researchers to assign the genetic linkage groups of potato to specific chromosomes, and the potato chromosomes were consequently numbered according to their correspondent linkage group (this numbering system is generally accepted and was also adopted by the International Potato Genome Sequence Consortium). Potato karyotypes based on pachytene chromosomes have been described and, as done in other species and for conventional karyotyping, chromosomes were numbered based on their descending length.

A recent treatment by Peloquin et al. described the pachytene-based karyotype of the *S. andigena* haploid, which consisted of three chromosomes with median primary constrictions, three with submedian constrictions, five with subterminal constrictions and one with a terminal constriction. The amount of euchromatin of each chromosome varied from 21 to 75%. Although the distribution patterns of euchromatin and heterochromatin and the differential staining of the centromeric regions enabled researchers to distinguish each chromosome, pachytene analysis and chromosome identification remains very difficult in potato.

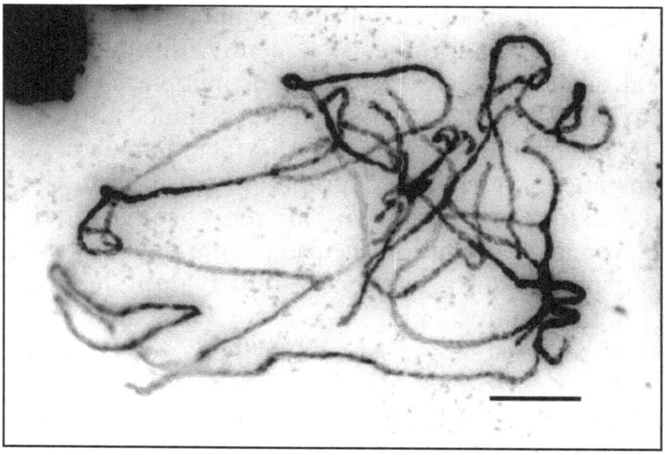

Fig. Pachytene chromosomes of a S. tuberosum haploid (2n = 2x = 24)

The genetics of the tetraploid potato is very unique and more complicated than that of either diploid crops or polyploids with disomic inheritance. At a given locus, five genotypes (aaaa = nulliplex, Aaaa = simplex, AAaa = duplex, AAAa = triplex and AAAA = quadruplex) are possible. In reality, there could be as many as four alleles per locus, providing the opportunity for rich interaction patterns. In polysomic polyploids like the potato, the more diverse the alleles are within a locus, the higher the heterozygosity and the greater the number of increased interlocus and epistatic interactions.

While diploids hold only one non-additive genetic effect at a locus, in the tetraploid potato there may be three non-additive genetic effects: first order, second order and third order. In a tetrallelic genotype this corresponds to 11 heteroallelic interactions. A unique advantage for breeders is that genotypes displaying tetrallelic genic interactions can be immediately fixed through clonal propagation. For traits whose genetic variance is almost entirely non-additive, this is a very important issue, for the breeder must strive to obtain maximum heterozygosity to optimize heterotic combinations. Indeed, in the tetraploid potato a drastic decline in yield and fertility occurs upon self-fertilization and in progenies from closely related parents. This can be attributed not only to increased homozygosity but also to the loss of favorable allelic intralocus interactions and of favorable epistatic combinations required to maximize heterosis. In designing breeding schemes aimed at maximizing heterozygosity, breeders need to be very careful in the appropriate use of parents. High levels of heterozygosity can be reached when new genetic materials are synthetized through controlled crosses involving genetically diverse parents or gametes with unreduced chromosome numbers, also referred as 2n gametes.

Multiple allelism in potato has been demonstrated for genes affecting either metabolic pathways or morphological traits. van Eck et al. reported the presence of three alleles at the *Ro* locus controlling tuber shape. van de Wal et al. found at least eight different alleles coding for granule-bound starch synthase I, responsible for amylose biosynthesis. Recently, through SNPs genotyping, Achenbach et al. demonstrated the existence of multiple alleles either at a single locus or at various linked loci controlling resistance to *Globodera pallida*. Complexity to inhertance is added by the fact that, depending on gene-centromere distance, alternative segregation patterns in the gametes are possible: chromosome segregation, random chromatid segregation and maximum equational segregation. Chromosome segregation occurs when chromatids derived from a particular multivalent belong to different chromosomes of that multivalent.

This type of segregation is assumed for genes completely linked to the centromere. In this scenario, the duplex genotype AAaa, forms AA, Aa and aa gametes in a 1:4:1 ratio. If either random chromatid segregation or maximum equational segregation occurs, aa gametes derived from sister chromatids are also expected. This phenomenon has been termed double reduction, characterized by a coefficient of double reduction ($\propto$) defined as the probability of two sister chromatids ending up in the same gamete. The cytological bases for double reduction include multivalent pairing with one cross over event between the locus and the centromere and the two pairs of chromatids from that cross over event passing to the same pole in anaphase I, and random separation of chromatids in anaphase II. The theoretical extreme values for μ generally reported in the literature are: 0 for chromosome segregation, 1/6 for maximum equational segregation and 1/7 for random chromatid segregation.

If the same duplex genotype AAaa mentioned before is considered, AA, Aa and aa gametes are formed in a 2:5:2 ratio with maximum equational segregation, in a 3:8:3 ratio with random chromatid segregation if $\propto = 0$, and in a $1+2\propto/6$: $4\text{-}4\propto/6$: $1 + 2\propto/6$ ratio with both types of chromatid segregation if μ is different from 0. Some other genetic features that are important in this context should be underlined. As with every clonally propagating crop, the cultivated potato is highly heterozygous. Consequently, segregation of traits is expected in the F1 generation following hybridization. Homozygous genotypes are difficult to obtain upon selfing due to inbreeding depression. The cultivated potato also has a narrow genetic base compared with that at its center of origin and with wild species. This is mainly due to the limited number of genotypes that were used to develop the *S. tuberosum* form of the cultivated potato.

REPRODUCTIVE CHARACTERISTICS RELEVANT FOR BREEDING

Production of 2n Gametes

A noteworthy reproductive characteristic with important potato breeding applications is that several genotypes of cultivated and wild potatoes produce *2n* gametes. They are the result of meiotic anomalies affecting various nuclear and cytoplasmic aspects of sporogenesis (chromosome pairing, centromere division, spindle formation or cytokinesis). For a review on meiotic variants leading to *2n* gamete formation in potato see Peloquin et al. and Carputo et al.. Three distinct meiotic mutations leading to *2n* pollen production have been described: parallel spindles (ps), including also fused and tripolar spindles, premature cytokinesis 1 (pc1), and premature cytokinesis 2 (pc2), all inherited as simple Mendelian recessive traits. When these mutations are present, at the end of meiosis dyads with two *2n* microspores are formed.

Watanabe and Peloquin reported that ps is the most common mutation leading to *2n* pollen production in the potato. Meiotic mutations affecting *2n* egg production have also been found and described. Werner and Peloquin reported that omission of second meiotic division (os) was the most common mutation of *2n* egg formation. When os is present, two *2n* megaspores are produced, one degenerating and the other yielding the functional spore. Another meiotic variant resulting in *2n* eggs is the failure of cytokinesis after the second division, which results in the fusion of nuclear products of this division. Other mutations leading to *2n* egg formation may affect megasporogenesis. These cause either an irregular anaphase II and delayed meiotic divisions or an irregular spindle axis formation at metaphase I. Each meiotic mutation leading to *2n* gametes is genetically equivalent to either first division restitution (FDR) or second division restitution (SDR) mechanisms.

It should be pointed out that genetic mode of *2n* gamete formation and cytological event are two separated entities. Thus, for example, ps occurs at second meiotic division but it is genetically equivalent to a FDR mechanism, whereas os (occurring during first meiotic division) is genetically equivalent to SDR. Due to the genetic consequences associated with the mode of *2n* gamete formation, various levels of variability, fitness, and heterozygosity are expected in newly formed polyploids. The genetic control of *2n* gamete production has been generally attributed to the action of single recessive genes. These genes exhibit incomplete penetrance and variable expressivity, and their phenotypic expression may be significantly modified by genetic, environmental, and developmental factors. Recently, the first gene implicated in the parallel orientation of spindles and $2n$ pollen production in *Arabidopsis thaliana* has been isolated and characterized at the molecular level. This gene encodes a protein of about 1,500 amino acids conserved throughout the plant kingdom, and it is a good candidate for the *ps* locus in potato.

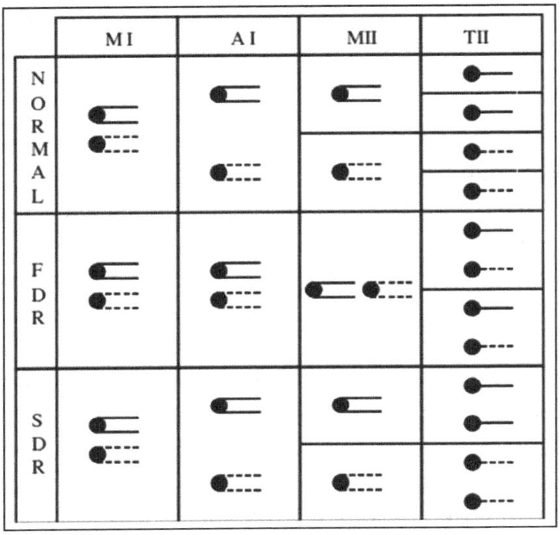

Fig. Representation of metaphase I (MI), anaphase I (AI), mataphase II (MII), and telophase II (TII) of a meiocyte with twoo homologous chromosomes undergoing: normal meiosis, first division restitution (FDR), and second division restitution (SDR) mechanisms.

Barriers to Interspecific Hybridization

There are three additional reproductive features of the potato that must be underlined: pollen-pistil incompatibility, nuclear-cytoplasmic male sterility, and endosperm development. They have important implications because there is strong and circumstantial evidence that, rather than chromosome differentiation, they provide effective barriers to interspecific hybridization at the pre-zygotic and/or post-zygotic level. Most diploid potato species are self-incompatible because they possess a single polymorphic *S* locus determining gametophytic incompatibility. This locus includes two genes, *S-RNAse* and *SLF/SFB*, controlling female and male specificity, respectively. Glycoproteins with ribonuclease activity control pistil specificity. By contrast, *S* gene products in pollen have not been identified so far. Genetic analysis suggested that a gene (*Sli*) outside the *S* locus may inhibit the self-incompatibility mechanism. Unilateral and bilateral cross incongruity may also occur. Camadro and Peloquin hypothesized a genetic model with dominant cross incompatibility genes in styles that prevent fertilization by pollen carrying specific dominant complementary genes. The model, which assumes segregation for both type of loci, accounts for the results of both unilateral and bilateral cross incompatibility in interand intra-specific crosses. A second reproductive characteristic that guarantees species integrity in tuber-bearing *Solanum* species is nuclear-cytoplasmic male sterility. This isolating barrier is connected with specific interactions between nuclearencoded proteins of one species and cytoplasmically-encoded proteins of another species.

As a result of this interaction, the progeny is male sterile. The existence of nuclear-cytoplasmic male sterility has been reported as a result of either interactions between *S. tuberosum* cytoplasm and nuclear genetic factors of wild species (e.g., *S. sanctae-rosae, S. infundibuliforme, S. raphanifolium, S. commersonii*) or species cytoplasm and nuclear genes of *S. tuberosum*. The importance of this type of sterility in Solanums was emphasized by Grun, who described a number of genes affecting anther indehiscence, sporad formation, anther style fusion, etc. Information about the genetic bases of nuclear-cytoplasmic male sterility is limited. Iwanaga et al. suggested the action of male fertility restorer genes. Ortiz reported that one or a few nuclear genes can interact in incompatible cytoplasms to cause male sterility. Molecular analysis in tuber-bearing *Solanum* species suggested that the nuclear-cytoplasmic male sterility is controlled by mitochondrial DNA. In potato, as well as in several other angiosperms, endosperm breakdown is considered the primary reason for seed failure after double fertilization. Genetically, the nuclear constitution of the endosperm differs from the embryo only in having one extra set of maternal chromosomes.

Thus, endosperm development is dependent on the same genes as the embryo but in different doses. To explain failures in interspecific crosses in *Solanum*, Johnston et al. developed a model that extended the 2:1 maternal to paternal genome dosage in the endosperm described in corn. This model relies on a balance of qualitative genetic factors and not genomes for normal endosperm development. In this model, each species has an effective number, which is not necessarily a direct reflection of its ploidy. It is the EBN which must be in a 2:1 maternal to paternal ratio in the hybrid endosperm for normal development of this tissue and, consequently, of the normal hybrid embryo. Thus, successful interspecific hybridization occurs only when parents produce gametes with the same EBN. The EBN is an arbitrary value experimentally assigned to each *Solanum* species based on its behavior in crosses with EBN standards, and assuming the 2:1 ratio as a prerequisite for normal endosperm development. Cultivated *S. tuberosum* has been assigned 4EBN. Diploid species have been assigned either 1EBN (e.g., those belonging to *Etuberosa* and *Commersoniana* series) or 2EBN (e.g., all diploid cultivated species and most wild species, like *S. multidissectum, S. chacoense, S. verrucosum*). Tetraploid species have been assigned either 2EBN or 4EBN. All the hexaploid species, such as *S. demissum* and *S. oplocense*, have been assigned 4EBN. Several genetic studies provided evidence that an oligogenic control of EBN exists in potato species.The essential role of EBN as an internal reproductive barrier to interspecific hybridization is that it creates effective barriers between sympatric species, leaving intact their genotypic integrity.

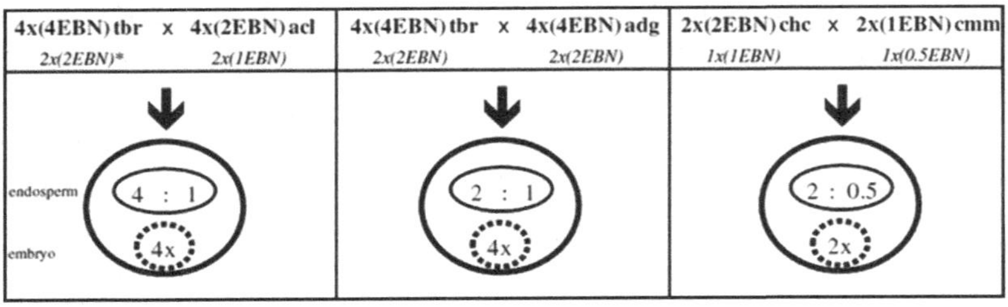

Fig. Prediction of female:male EBN ratio in the hybrid endoperm (solid oval) and ploidy level of the hybrid embryo (hatched circle) resulting from interspecific crosses involving Solanum species with different EBN. Tbr = S. tuberosum, acl = S. acaule, adg = S. andigena, chc = S. chacoense, cmm = S. commersonii. Only cross combinations resulting in a 2:1 female to male EBN ratio [e.g., 4x(4EBN) tbr × 4x(EBN) adg] in the hybrid endosperm are compatible. *Gamete ploidy (EBN).

For this reason, for example, 2x *S. commersonii* cannot be crossed with *S. chacoense*, which is also diploid but with 2EBN. On the other hand, *S. commersonii* hybridizes with other diploid species. The EBN represents a powerful screen for 2*n* gametes during sexual polyploidization events that lead to the polyploid evolution of *Solanum* species. Inter-EBN crosses can occur, but they are only sporadic events that probably result from non-heritable random events, such as multiple fertilizations of the central cell, mitotic abnormalities in the gametophyte, endomitosis of the polar nuclei in the endosperm, or increase in the number of polar nuclei.

CONVENTIONAL BREEDING METHODS

Potato enjoys an array of crossing procedures that makes it a model crop for breeders. It is amenable to various in vitro culture techniques such as somatic embryo development, protoplast fusion, cell, anther and pollen culture. It is also an ideal species for *Agrobacterium*-mediated transformation protocols that allow the development of trans- and cis-genic genotypes. However, in this section only conventional breeding methods based on sexual hybridization will be considered. These can be classified into two levels based upon ploidy and EBN: (a) the tetraploid 4EBN level, and (b) the diploid 2EBN level.

THE TETRAPLOID 4EBN LEVEL

Tetraploid breeding is essentially based on phenotypic recurrent selection, involving crosses between 4*x* genotypes and then field evaluation and selection of descendants for several years. Most currently used varieties derive from this breeding approach. Its genetic and reproductive bases have been presented in previous sections of this chapter. The procedure starts with the selection of diverse, desirable 4*x* parents, a crucial aspect to prevent homozygosity and guarantee high allelic diversity of offspring. Parents are often selected based on their phenotype rather than genotype, and crosses are made between pairs with complementary characteristics. However, due to the non-additive gene action for several traits of interest, phenotypic parental selection is often ineffective. Therefore, the use of test-crosses to identify desirable combination of parents has been reported by various authors.

Test-crosses evaluate both general combining ability (GCA) and specific combining ability (SCA). The former refers to the average performance of a clone in crosses with other clones, the latter to the performance of a clone in a specific combination in comparison with its performance in other specific combinations. Reviews by Tarn et al. and Bradshaw and Mackay provide a detailed list of reports on combining ability analysis using *S. tuberosum* genotypes as parents.There are essentially two traditional approaches to calculate combining abilities, one based on crosses in all directions between two groups of complementary clones, and the other method based on diallel crosses. Recently, to reduce the number of matings normally used in test-crosses to estimate GCA, Gopal et al.

Proposed the use of top-crosses (matings with selected single testers). Tarn et al. suggested that when GCA is predominant, valuable cross combinations can be determined simply based on the GCA information of parents. When SCA is larger, breeders should carry out progeny tests on samples of progenies produced within the breeding program. It should be pointed out that superior parents may also derive from pre-breeding activities aimed at producing improved tetraploid parents.Due to the high levels of heterozygosity of tetraploid potatoes, segregation of traits is expected in the F1 generation derived from $4x$ x $4x$ crosses. In a typical program, the year after crossing large populations of F1 seedlings are grown for visual selection. A single tuber is usually taken from each selected F1 plant for planting in the next season in a single-hill plot. This represents the first clonal generation. During the next two to three seasons, clones are cultivated and screened in larger unreplicated plots with plants spaced for further evaluations and selection and to increase the number of seed tubers. After a number of years of such selection, superior clones are tested in replicated trials in different locations to estimate the genotype x environment interaction. In the breeding flow described, a high percentage of clones is discarded in the first clonal generations based on field evaluations. Resistance and quality traits are usually screened during the final stages of selection, when the number of clones under evaluation has been drastically reduced. At later stages, breeders generally select simultaneously for a very high number of traits. Tarn et al. outlined that while at later stages of selection emphasis is given to characteristics of individual clones, at early stages of selection parents and crosses play a significant role in breeders' decisions. For traits with low heritability, stringent visual selection at early generations is usually ineffective and increases the risk of losing valuable clones. To raise the efficiency of early generation selection, various authors have developed modified selection procedures. Gopal et al. suggested elimination of seedlings with poor vigor before transplanting; the rejection of clones with undesired color, shape, eye depth and tuber cracking at the seedling stage; negative selection for tuber yield and weight at the first clonal generation and for low tuber number at the second clonal generation. Negative selection involves the elimination of clones with poor traits and the retention of those with no obvious serious defects.

Bradshaw et al. proposed the combined use of mid-parent values to select the best parents and progeny tests to discard whole progenies before starting the within progeny selection. They also suggested increasing the number of clones on which to practice selection by sowing additional seeds from best progenies.

Among several other reports outlining strategies of early generation selection for important traits, are those by Thill and Peloquin for chipping ability, Bisognin and Douches for tuber appearance, specific gravity and chip color, and Bae et al. for resistance to Verticillium wilt. In sum, classical breeding at the tetraploid level is not very different in principle from that performed at the beginning of the 20th century. Its rationale is essentially based on the evaluation of a decreasing number of selected clones in increasingly elaborate field/ laboratory trials in a wide range of environments. The last part of the breeding program includes official regional variety testing, healthy maintenance and propagation of seed tubers and marketing strategies.

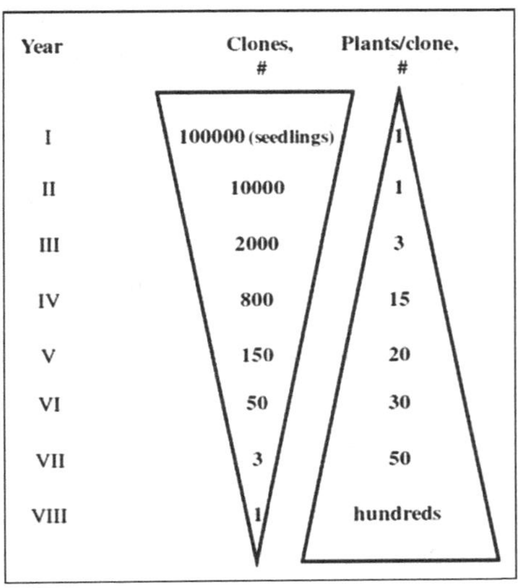

Fig. Schematic representation of a traditional potato breeding scheme. A hypothetical number of clones are evaluated each year and a hypothetical number of plants are grown for each clone. Following crosses between selected parental lines, each year a decreasing number of selected clones is evaluated in increeasingly sophisticated field/ laboratory trials.

THE DIPLOID 2EBN LEVEL

In modern potato breeding, several programs operate at least in part at the diploid level. This not only takes advantage of disomic inheritance, which is less complicated than the typical tetrasomic patterns of the 4*x* level, but also requires a smaller population size at the start.

In addition, this approach allows the most efficient exploitation of diploid potato germplasm. The first essential ingredient for potato breeding at the diploid 2EBN level is haploids (sporophytes with the gametic chromosome number). *S. tuberosum* maternal haploids can be easily obtained following crosses with diploid clones of *S. phureja*.

Starting from the pioneering work of Hougas and Peloquin, *S. tuberosum* haploids have been routinely produced for breeding purposes, and also for genetic, cytogenetic and evolutionary studies.

It is believed that haploids originate via parthenogenesis, and that both sperm nuclei fuse with the polar nuclei of the central cell of the female gametophyte. Clulow et al. hypothesized that *S. tuberosum* haploids may also originate from a process of chromosome elimination from zygotes deriving from *S. tuberosum* x *S. phureja* crosses.

Wilkinson et al. and Ercolano et al. demonstrated stable incorporation of *S. phureja* DNA into haploids obtained through crosses with *S. phureja*. Potato haploids may also be obtained through in vitro microspore/anther culture, but this method is more complicated and strongly genotype-dependent compared to the *S. tuberosum* x *S. phureja* approach. From the breeding standpoint, *S. tuberosum* haploids are widely used in sexual interspecific hybridization programs.

They often represent essential ingredients also for the production of somatic hybrids through protoplast fusion. Since potato haploids have a 24-chromosome complement and an EBN = 2, they are essential tools to efficiently capture the genetic diversity of 2*x* *Solanum* species through haploid x wild species crosses.

In addition, they also allow breeders to indirectly evaluate tuber characteristics of wild species (several potato species do not tuberize under long-day conditions).

It should be pointed out that haploid-wild species hybrids have 50% wild genome. Therefore, after population improvement at the 2*x* level for traits of interest, further breeding is needed to re-establish the chromosome number of the tetraploid cultivated potato and gradually reduce the wild genome content.

This is accomplished through 2*n* gametes, representing the second essential feature of breeding at the diploid 2EBN level. Haploid-wild species hybrids producing 2*n* gametes and with the desired traits are selected for the production of tetraploid hybrids through bilateral and unilateral sexual polyploidization.

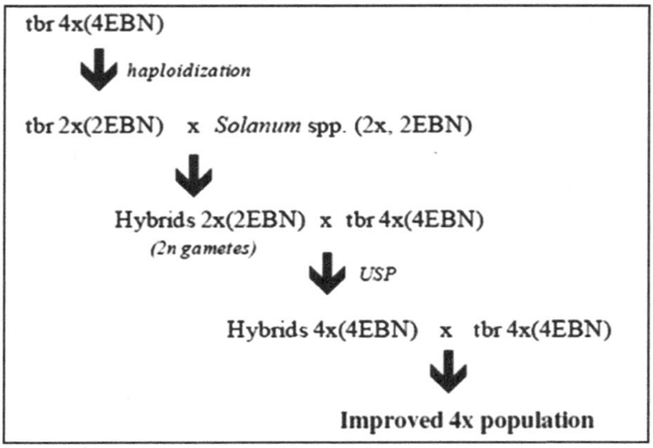

Fig. Breeding at the diploid (2EBN) level involving the production of Solanum tuberosum (tbr) haploids and the selection of superior 2x(2EBN) interspecific hybrids producing 2n gametes. In this outline, the unilateral sexual polypolidization crossing scheme (USP) is employed.

BSP involves crosses in which the female parent forms *2n* eggs and the male one *2n* pollen. By contrast, USP is based on crosses where only one parent produces *2n* gametes.

The exploitation of *2n* gametes and the asexual reproduction system of the potato provide the breeders the unique possibility of accumulating non additive genetic effects at the tetraploid level of the offspring and the immediate fixation of heterosis, respectively.

The extent of allelic interaction transmission from haploid-species hybrids to the tetraploid progeny depends mainly on the mode of *2n* gamete production, whereas the creation of new interactions depends mainly on the allelic diversity of parents involved.

It has been estimated that with FDR all loci from the centromere to the first crossover that are heterozygous in the parent will be heterozygous in the gamete, and one half of heterozygous parental loci beyond the first chiasma will be heterozygous in the gametes. By contrast, with SDR mechanisms, all the loci from the centromere to the first chiasma will be homozygous in the gametes, and all loci past the first crossover, which are heterozygous in the parent, will be heterozygous in the gametes.

Based on the positions of centromeres and chiasmata and on the fact that in the potato generally only one chiasma per chromosome arm occurs, it has been estimated that the percent of heterozygosity transmitted by gametes is about 80% with FDR and 40% with SDR. FDR gametes are expected to strongly resemble each other and the parental clone from which they originate.

They also maintain and transmit a large amount of inter-locus epistatic interactions, which are important in maximizing heterosis. By contrast, SDR is expected to produce a heterogeneous population of highly homozygous gametes.

The choice of USP vs. BSP depends on the mode of $2n$ gamete formation in the parent and, in the case of USP, on the fertility of the 4x parents.

Due to the high level of allelic interactions transmitted by FDR gametes and on the wide occurrence of $2n$ pollen, the use of the $4x$ x $2x$-FDR approach has been documented thoroughly. Ortiz et al.

Reported that besides the allelic diversity transmitted, the $4x$ x $2x$ breeding scheme is more effective than the traditional $4x$ x $4x$ scheme also because fewer replications and locations are necessary for progeny evaluation. Another advantage of the $4x$ x $2x$ breeding scheme is that it can increase the efficiency of cultivar development. The authors found that the percentage of clones combining high yield and good tuber quality traits selected in relatively small $4x$ x $2x$ populations was much higher than that reported using $4x$ x $4x$ crosses.

BSP can be employed as an alternative to USP when superior $2n$ egg and $2n$ pollen producing $2x$ parents are available. This method potentially allows the combination of multiple traits from two diploid interspecific hybrids into tetraploid genotypes through $2x$ x $2x$ crosses.

It also allows transmission of high levels of heterozygosity into $4x$ progeny, especially when $2x$ hybrids used have unrelated genetic backgrounds. Werner and Peloquin reported a high increase in tuber yield of $4x$ families from $2x$ x $2x$ crosses in comparison to control cultivars and advanced selections.

In $2x$ crosses, SDR and FDR complement each other in terms of transmission of heterozygosity. Species exploited in potato breeding programs at the diploid 2EBN level include, among others, *S. berthaultii, S. tarijense, S. bukasovii, S. sparsipilum, S. phureja, S. multidissectum,* and *S. chacoense.*

In Poland, Scotland, Canada, China, the USA and other countries with potato breeding programs, haploid–species hybrids with multiple resistances and tuber quality are developed and used in $4x$ x $2x$ crossing schemes to produce parental lines and cultivars.

It should be pointed out that crossing barriers sometimes do not allow the exploitation of breeding at the diploid 2EBN level. For example, $2x$ species cannot be hybridized with $2x$ *S. tuberosum* haploids. Similarly, crosses between $4x$ species and $2x$ potato haploids produce triploid offsprings.

In these cases, breeding schemes based on ploidy manipulations and bridge crosses can be designed. These result in intermediate hybrids that allow the successful exploitation of potato species.

Figure illustrates the breeding schemes used by Carputo et al. to transfer useful genes and genetic diversity from $2x$ *S. commersonii* to *S. tuberosum*. Additional strategies to overcome hybridization barriers in potato include mentor pollinations and embryo rescue, hormone treatment, reciprocal crosses, careful selection of parents and, of course, somatic hybridization.

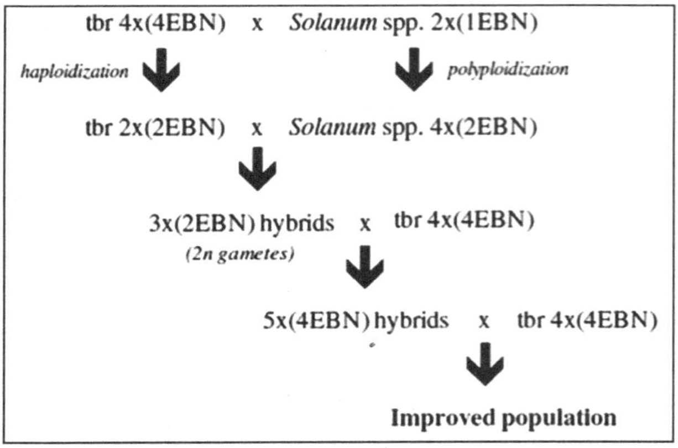

Fig. Breeding scheme aimed at the exploitation of incongruent 2x(1EBN) Solanum species. Ploidy manipulations are necessary to make parental EBNs compatible. Triploid and pentaploid bridge ploidies are produced. Polyploidization of the 2x(1EBN) species can be performed somatically through either antimitotic metabolites (e.g., colchicine) or in vitro spontaneous chromosome doubling. Alternatively, in a 2n gamete producer, the 2x(1EBN) species can be directly crossed with tbr 2x(2EBN) haploids.

BREEDING OBJECTIVES

Recent world issues concerning environment protection, agriculture sustainability, food safety and quality have largely impacted potato breeding objectives. Furthermore, as recently outlined by Bonnel, two additional aspects have affected breeders' activities: the increase in trade exchanges between countries traditionally involved in potato production and the rapid emergence of new countries that have become leaders in potato production. In light of this situation, breeders are focusing more and more on the development of regional or international varieties rather than local or national ones. At the same time, however, markets with specific requirements are also emerging. This is the case of organic potato production. Traditionally, breeders consider more than 50 traits during field and laboratory evaluation and selection.

These can be grouped into three categories:

- Yield;
- Quality characteristics of tubers; and
- Resistance to biotic and abiotic environmental constraints.

Quality has become a challenging target for potato breeders worldwide. Quality parameters can be organized into two major categories. The first category includes external characteristics, comprising skin color, tuber size and shape, and eye depth. These traits are deemed very important for fresh consumption, where they are most likely to influence consumer's choice. The second category comprises internal characteristics, including nutritional properties, cooking/after cooking properties, and processing quality.

Internal quality encompasses traits such as dry matter content, flavor, sugar and protein content, starch quality, type and amount of glycoalkaloids. There are several factors affecting tuber quality: genotype, earliness, field practices, environmental conditions, storage temperatures, and the impact of pests and diseases. Of these, genotype is probably the most important. Traits that are genetically controlled include biological (proteins, carbohydrates, vitamins, minerals, reduced amounts of toxic glycoalkaloids); sensorial; and industrial (tuber shape and size, dry matter content, cold sweetening, oil absorption, starch quality) traits. Breeding objectives related to quality for processed potatoes are normally different from those for fresh use. The former are well defined and are mainly related to dry matter content, level of reducing sugars during cold storage, starch type and, to some extent, tuber shape and size.

The latter fall more into the category of external traits, even though eating quality traits such as after-cooking blackening are also important. Renatus pointed out that breeders need to consider two main types of potatoes for the fresh market. The first includes the standard merchandise at a discount price level, where breeders are required to provide cost efficient, high yielding, multi-use varieties. The second category comprises nice-looking, high price, top quality potatoes. This high-end segment needs the release of varieties with superior taste, increased nutritional value, and superior appearance.

Among new goals of potato breeding for quality, is the development of varieties that do not accumulate asparagine. This aminoacid is the precursor of acrylamide, formed during the Maillard reaction as part of the cooking process. It may cause damage to the nervous system and has carcinogenic effects. Equally important is the need to control tuber dormancy, in order to avoid the use of expensive chemical suppressants that leave chemical residues, and the identification of varieties with low glycoalkaloid accumulation in the tubers. It is widely known that potato glycoalkaloids are extremely toxic to humans and animals, having possible teratogenic, embryotoxic, genotoxic, and carcinogenic effects. In terms of biotic stress resistance, of primary importance is *Phytophthora infestans,* the causal agent of late blight.

Since its first outbreak in Europe during the 19th century, late blight is considered the most devastating fungal disease of potato, causing decay of both foliage and tubers. Race-specific resistance has been introgressed into the cultivated gene pool from wild *S. demissum*. However, new *P. infestans* races have overcome it. In recent years, breeding activities on late blight resistance have been brought together by the Global Initiative on Late Blight, a network of researchers and technology developers from all over the world. In recent years, resistance to various pathogens and pests (e.g., nematodes, *Colletotrichum coccodes, Leptinotarsa decemlineata, Streptomyces scabies, Rizochtonia solani, Verticillium* spp., *Clavibacter michiganensis* and *Erwinia chrysantemii*) has become a major objective for potato breeding.

Despite the very high number of other biotic constraints that may hamper potato cultivation, not all pathogens and pests are present in all areas where potato is grown. Resistance to nematodes, for example, is an important breeding target in Europe but not in the USA. In Europe, the potato is a major host for potato cyst nematodes *Globodera pallida* and *G. rostochiensis* and for root-knot nematodes. Here, the availability of varieties resistant to *G. rostochiensis* has accentuated the presence of *G. pallida*, and therefore breeding efforts are being directed towards the development of genotypes resistant to this cyst nematode.

Among abiotic stresses, resistance to high temperatures and water stress is deemed very important due to the trend of increased potato cultivation in warm climates. It should be pointed out that in these areas cultivation is carried out using varieties that were developed for cool climates. Also noteworthy is resistance to low temperatures in environments where the potato can be grown all year around. In the Mediterranean region, for example, the potato is planted starting in November, in a cycle that is much earlier than the typical spring-summer cycle. Due to the lack of cold resistant varieties, this off-season production can be seriously damaged by frost events. Specific objectives are required in potato programs aimed at producing true potato seed (TPS).

TPS technology represents a possible alternative to the traditional use of seed tubers. It has several advantages over the traditional cultivation, including a reduction in production/management costs and in the incidence of diseases. TPS varieties represent a collection of genotypes deriving from a cross between two heterozygous parents that produce a phenotipically uniform progeny.

Therefore, in a typical TPS breeding program, the selection of parental lines is an important prerequisite. Both GCA and SCA may be important in parental source identification. TPS parents must guarantee satisfactory botanical seed production when crossed. At the same time, they must result in progenies with outstanding tuber yield and traits. Therefore, along with breeding objectives common with other production systems (e.g., tuber yield and quality, resistances), specific objectives related to flower duration and intensity, number of flowers per inflorescence, vigorous early growth of seedlings, emergence uniformity, etc. are also important. In other words, antagonism should not occur between tuber growth from one side and viable botanical seed production from the other.

MOLECULAR BREEDING FOR POTATO IMPROVEMENT

MOLECULAR BREEDING—THE BEGINNING OF A NEW ERA

We have entered a rapidly changing period where new technology and new analysis techniques provide opportunities for crop improvement. Molecular markers have long held the promise of improving the efficiency and effectiveness of plant breeding. Due to the autotetraploid state of potato, researchers have primarily utilized diploid genetic materials for genetic map construction and the identification of markers linked to important traits. These markers have proven useful for enhancing our understanding of cultivated potato and wild *Solanum* species used as sources of beneficial diversity by potato breeders. Application of association genetics to tetraploid breeding populations will further contribute to the identification of valuable markers for potato improvement. Gene cloning and transformation technology offer a route to overcome the weaknesses inherent to widely grown potato cultivars; however, lack of consumer and industry acceptance of genetically modified organisms has limited the impact of this technology to date.

Advances in potato, and the closely related tomato, genome sequencing will greatly improve our knowledge of the gene content and genome organization of potato. Next-generation sequencing, high-throughput genotyping, and bioinformatics will further contribute to the evaluation of allelic diversity existing in *S. tuberosum* and related species. The question that has yet to be answered is whether this wealth of genomics information will have an impact on potato breeding. The goal of this chapter is to develop realistic expectations for molecular breeding and to place molecular breeding in the proper context within a larger potato improvement program. Fundamental to determining the role of markers for plant improvement is an understanding of the basic principles of selection.

The three main principles are:

- Selection operates on existing variation;
- Selection acts only through heritable differences; and
- Selection promotes reproduction of some individuals at the expense of others.

The use of markers will only be useful in crop improvement when the target traits have these selection criteria. It should also be obvious then that there are breeding targets for which markers will have little or no impact—either due to a lack of understanding of the trait, a lack of heritable variation for the trait, or a negative relationship between the target trait and other critical traits. In some cases, the role of transgenics-assisted breeding should be considered as a more viable option for meeting breeding goals. Molecular breeding also needs to be viewed in a crop-dependent manner. Potato has unique cultural requirements, reproduction methods, and breeding strategies. Therefore, this chapter will take a potato-orientated approach and review issues that may be more important for potato than for other crops. We will review the application of genomics technology within potato germplasm evaluation and breeding programs and identify new opportunities. This chapter is written from the perspective of a scientist within a tetraploid potato improvement program in a "developed" country with a long-day growing season.

GERMPLASM AND VARIETY CHARACTERIZATION

DNA marker and sequencing information is playing a significant role in taxonomic studies. While taxonomic investigations do not have an obvious relationship to breeding, enhanced understanding of gene pools available to geneticists and breeders will lead to better utilization of these resources. Random amplified polymorphic DNA, amplified fragment length polymorphism and simple sequence repeat (SSR) markers have been used to generate phylogenetic data that have contributed to the revision of the relationships within cultivated and wild potato species germplasm. Two topics of particular interest to potato breeders are the phylogenetic relationships within the primary and secondary gene pools of cultivated potato and the elucidation of the progenitors of cultivated *S. tuberosum*.

Characterization of **Solanum** *Gene Pools*

Molecular data, generated by sequencing or DNA markers, has generally led to fewer unique species or taxonomic groups compared to the initial morphology-based studies summarized by Hawkes. Currently, there are 187 recognized tuber-bearing species. Within *Solanum* section *Petota*, the grouping of species into 21 different series was proposed. Molecular evidence does not support this number, instead supporting between four and 10 clades. Various types of markers have contributed to the classification changes within section *Petota*, including AFLPs, SSRs, or a combination of AFLPs, RAPDs, and SSRs.

More recently, sequencing data from single-copy genes has been used to elucidate the relationships within *Solanum*. Genes that have been studied include *GBSSI* and nitrate reductase. This new molecular evidence has been particularly informative in studying polyploids within the genus. Polyploidy is thought to have contributed to the expansion of the range of species in the genus. Of the 189 potato species, 64% are diploid, and 36% are partially or completely polyploid. The importance of polyploidy within this taxonomic group was illustrated in this study by the amount of geographic area containing only polyploid species (28%) and that the two most widespread species are tetraploids. Two important aspects of polyploidy within *Solanum* section *Petota* revealed by the study of single gene sequence data include molecular evidence for allopolyploidy and the lower level of diversity present in diploid versus polyploidy species.

These authors also examined the relationship between genome type and endosperm balance number (EBN). Potato species have been typically classified by the type of genome they possess based on meiotic pairing studies of interspecific hybrids. The classifications include A-E and P, with *S. tuberosum* being an A genome species. While there isn't a perfect relationship between EBN and genome type, it is apparent that genome-specific markers would allow breeders to more effectively access diversity within non-A genome species. In the Rodriguez and Spooner and Spooner et al. studies, the level of diversity within a species was significantly higher in polyploids compared to diploids. These polymorphisms were in the form of single nucleotide differences and insertion/deletions. The issue of ploidy level is also very important in the study of cultivated potatoes and requires a critical assessment of the choice of marker platform for these studies.

Genealogy and Diversity of Modern Varieties

Molecular markers have been utilized to help unravel the mystery of the origin of the modern cultivated potato. The hypothesis of rapid adaptation of short-day Andigena types of *S. tuberosum* to the long-day environment of Europe has been challenged recently by proponents of the introduction of *S. tuberosum* ssp. types from Chile. The molecular evidence strongly indicates that modern cultivars arose from a Chilean ancestor rather than from Andigena.

The implications of this paradigm shift include a re-evaluation of past and future uses of Andigena materials in long-day adapted breeding programs. The questionable origin of "Neo-tuberosum" materials generally suggests that potato pedigree information is often unreliable and that the negative selection pressure against Andigena type potatoes (short-day adaptation, deep eyes, etc.) needs to be factored into breeding schemes designed to enhance diversity. The study by Ghislain et al. indicates that modern varieties and breeding lines possess considerable diversity (93.8% SSR alleles in common with landraces). In a series of publications, cultivated accessions and modern varieties have been characterized using RAPDs, RAPDs and SSRs, SSRs, and AFLPs.

Comparison of AFLP, RAPD, and SSR markers for diversity analysis of tetraploid potato has not revealed a distinct advantage of one marker type over another. Standardized diversity analysis within cultivated potato species has been aided by the development of a microsatellite genetic identity kit by researchers at the International Potato Center. The various studies examining molecular diversity in cultivated potato have not demonstrated that diversity is a limiting factor for breeding progress. However, the high levels of polymorphism at the sequence level are generally not captured by the commonly used marker systems. These markers are also not sensitive enough to quantify allele dosage on a routine basis. Rather than marker assessments of genetic relationships and diversity, it may be more useful to measure the diversity of genomic regions that have undergone selection within breeding programs.

Distinctiveness, Uniformity, and Stability

Genome-wide marker evaluation can also be used to identify redundancy within a germplasm collection or conversely be used as a measure of distinctiveness of cultivars prior to commercialization. Published studies have documented duplication within potato genebanks and it is expected that markers will become more important in prioritization of accessions. Ghislain et al. illustrated the importance of utilizing more than one marker system for the establishment of core collections. Several researchers have documented the utility of molecular markers for discriminating potato cultivars using RAPD or SSR markers. The point at which genetic distinctiveness can be ascertained to declare a variety essentially derived is unclear. An assessment of genetic variability will need to be combined with morphological assessments, since the epigenetic variation plays a role in controlling phenotypic variation is now a topic of great interest in model plant species. AFLP and SSR markers failed to discriminate between intraclonal variants of "Russet Norkotah" and RAPD markers failed to distinguish intraclonal variants of four varieties, suggesting that epigenetic variation occurs in *S. tuberosum*.

MARKER-ASSISTED GENE INTROGRESSION

There is a growing collection of markers linked to important traits in potato. The genome location of pathogen resistance and tuber traits has been used to create potato function maps for pathogen resistance and tuber traits. An update of the location of important genes is provided in other chapters of this book. In crops it is generally viewed that markers have a role in both positive and negative selection within a germplasm enhancement program. These approaches in potato were reviewed by Barone.

Genome-wide Markers for Negative Selection

Selection against the donor genome is generally considered for pre-breeding applications in combination with phenotypic selection for a novel trait derived from a species distantly related to *S. tuberosum*. The strategy of introgression depends on the extent the donor species is reproductively isolated from the cultivated potato gene pool. Strategies for overcoming crossing barriers were reviewed by Jansky and include somatic hybridization and the use of bridge species. Once an initial hybrid plant is produced, backcrossing is typically employed as the breeding method. However, it is not uncommon to observe severe sterility or incompatibility in the early generation hybrids. When fertile hybrids are produced, genome-wide marker genotyping may be employed to preferentially select clones with a higher level of *tuberosum*-type markers, as long as the selected clones maintain the desirable trait. Most published examples of this marker-assisted selection (MAS) scheme in potato involve the diploid wild donor species *S. commersonii* introgression into a tetraploid *S. tuberosum* background.

The Toolbox of Available Markers

Positive selection with DNA markers has been more widely published, involving markers linked to several different traits. PCR markers closely linked to genes or quantitative trait loci (QTL) of interest have been available to potato breeders for quite some time. These include markers linked to nematode resistance genes, including *Gro1*, *Gro1-4*, *H1*, and *Grp1* genes for resistance to golden cyst nematode. Several PCR markers linked to virus resistance genes have been reported and include the *Ryadg* and *Naadg* genes on chromosome XI for extreme resistance to Potato virus Y (PVY) resistance and hypersensitive resistance to Potato virus A; the *Rysto* gene for extreme resistance to PVY; the *Rx1* gene for Potato virus X resistance; and the *Ns* gene for Potato virus S resistance.

PCR markers linked to oomycete and fungal resistance include the late blight resistance genes *RB* and *R1,* and QTLs; the *Verticillium* wilt resistance genes *StVe* and *Vc*; and the potato wart resistance gene *Sen1.* Several PCR-based candidate gene markers are associated with the tuber quality traits chip color and starch content and potato tuber skin and flesh color.

Marker-assisted Selection in Potato

There have been relatively few studies documenting the use of MAS within potato breeding programs. This section will summarize the published work on this topic and mention some recent findings that will contribute to future MAS studies. The markers linked to *Ryadg* and *Rysto* have been evaluated by several groups with varying degrees of success. The *Ryadg* marker genotypes were in good agreement with phenotypic assessment of extreme resistance to PVY when the source of resistance is known. In a recent study, Ottoman et al. validated the previously published markers for *Ryadg* by showing a 96.4% concordance with ELISA results for PVY in a full-sib tetraploid breeding population.

The authors also proposed a breeding scheme to incorporate the markers for selection after the first year of field evaluation. The pedigree of the material used to identify the initial *Rysto* markers on chromosome XI has been questioned. It is now generally recognized that *Rysto* is located on chromosome XII. However, the previously published markers were of limited use in a tetraploid breeding population, necessitating the screening of RAPD markers and development of sequence-characterized amplified region markers linked to the gene.

MAS for the *RB* gene in 110 breeding lines from different field generations was highly effective in selecting late blight resistance and less error-prone than detached leaf assays. Bormann et al. discussed a preliminary MAS study for late blight resistance based on three unlinked markers which resulted in a group of plants more resistant than a control group with unfavorable alleles at the same loci. They also reported a much higher frequency of susceptible alleles, which would necessitate much larger populations to identify sufficient progeny with the target alleles to allow for selection on other traits. For autotetraploid mapping and MAS experiments, Bradshaw and colleagues recommended the use of a minimum of 250 progeny per family. These observations could also be viewed as evidence for the use of MAS for parental selection to increase the frequency of desirable alleles segregating in a population. The genome region containing the *R1* locus showed a significant association with late blight resistance and plant maturity across a diverse set of germplasm and the markers could be traced back to the donor *S. demissum.* This important finding suggests that markers in this region may also contribute to developing lines with the appropriate phenotype.

An association genetic study with 184 clones from two breeding programs identified nine single nucleotide polymorphisms significantly associated with field resistance to late blight. The allene oxide synthase 2 gene on chromosome XI appears to be important in controlling this trait and markers targeting the diagnostic SNPs may be useful for future MAS applications. A diagnostic marker for resistance to *G. pallida* pathotypes 2 and 3 has been developed and holds promise for MAS. Five sequence tagged site (STS) markers tightly linked to the *Rmc(blb)* gene perfectly co-segregated with root-knot nematode resistance and also would be very suitable for use in a MAS program.

The predictive value of a marker associated with *Verticillium* wilt susceptibility was relatively high, but not high enough for MAS alone. A microsatellite allele was tested on 139 breeding clones and shown to be significantly associated with resistance to *Verticillium* wilt. This assay would be suitable for MAS, provided that maturity data are included in the selection decision. Multi-trait MAS was applied to select breeding lines with combined resistance alleles for *Ryadg*, *Rx1*, and *Gro1* or *Ryadg* and *Sen1.* The results from this study were better than expected, with selection of progeny containing resistance to PVY, PVX, and golden nematode, and unexpected resistance to *S. endobioticum* pathotypes 1, 2, and 6. In addition, twice as many clones with three target alleles were selected.

These unusual findings were explained by the use of parental clones which were tetraploid, not diploid as assumed, so the crosses were in fact $2x$ by $4x$ not $2x$ by $2x$. The contribution of wart resistance from the parents with extreme resistance to PVY explained the unexpected wart resistance. This study also supports the importance of parent development in potato molecular breeding, since parents producing $2n$ gametes transmitted the *Gro1* allele to 66% of the progeny. The use of tetraploid parents with more than one copy of a favorable allele would also significantly increase the proportion of progeny with the target traits. To increase the efficiency of MAS, Witek et al. developed a multiplex PCR assay that successfully identified the *Rysto* and *Ns* alleles for resistance to PVY and PVS, respectively. For tuber quality characteristics, there are fewer applications of markers within breeding programs. Several marker-trait associations were detected with chip color and starch content in 243 clones using 36 microsatellite and candidate gene loci. This study also validated the previous associations of the invertase alleles *InvGE-6f* and *InvGF-4b* with chip color in independent breeding lines. Tuber skin color is a trait controlled by three genes and can be considered a model trait to examine the role of candidate gene markers within a potato breeding program. A cleaved amplified polymorphic sequence marker derived from the anthocyanin pathway gene dihydroflavonol 4-reductase (*dfr*) was present in all red-skinned clones. Polymorphisms present in the *dfr* allele associated with red color were used to develop a dosage sensitive Taqman assay for genotyping tetraploid potato lines. Recently, a high-resolution DNA melting (HRM) probe assay was developed for the *dfr* gene and this assay was able to quantify the number of copies of the haplotype associated with red pigment and identify novel haplotypes. A unique allele was present in all chipping clones derived from "Lenape". It is also interesting to note that the *dfr* gene is located in close proximity to the sucrose synthase III gene on chromosome II, a gene that was found to be associated with light chip color in Canadian breeding lines.

Advances in Marker Technology for Autotetraploids

DNA marker systems developed for diploid organisms are not wellsuited for applications with autopolyploid organisms. The typical marker considerations for a MAS project include ease of use, robustness, cost, linkage to genes controlling the trait of interest, and the amount of phenotypic variation explained by the marker. In autotetraploid potato, a single genomic region may have up to four alleles or several possible combinations involving two or three alleles.

Therefore, dosage-sensitivity and the ability to distinguish various haplotype combinations and their dosage should be considered when selecting a marker system for potato molecular breeding. To address the limitations of electrophoresis-based marker systems, new SNP genotyping methods have been developed and applied to tetraploid potato. SNPs have been genotyped in tetraploids using pyrosequencing, dideoxy sequencing, and single nucleotide primer extension, fluorogenic 5' nuclease assays, ion-pair reversed-phase high-performance liquid chromatographyelectrospray ionization mass spectrometry, and allele-specific oligonucleotide microarrays. While these genotyping systems look promising, it is not apparent how they could be incorporated into a potato breeding program without considerable investment. The short analysis time, lack of post-PCR sample processing, and ability to quantify allele dosage indicate that HRM, reviewed by Erali, would be appropriate for MAS. Recently, we demonstrated that high-resolution DNA melting can be applied to genotyping and variant scanning of both diploid and tetraploid potatoes. One of the assays described in this study targets a haplotype of a marker linked to the *H1* gene. This HRM probe assay is now routinely being used to select breeding lines with pathotype *Ro1* golden nematode resistance within the Agriculture Canada potato breeding program in New Brunswick.

BREEDING STRATEGIES INCORPORATING DNA MARKERS

The conventional view of marker-assisted selection needs to be evaluated within the context of the steps in a potato breeding program. It is difficult to generalize, since each potato breeding program is unique; however, there are common elements to all programs and these will be emphasized. Based on the review presented in previous sections of this chapter, application of molecular breeding with DNA markers will be discussed pertaining to pre-breeding, parent development, early generation selection, and advanced generation selection.

Pre-breeding with DNA Markers

Marker-assisted germplasm characterization could occur on the genomewide or candidate gene level. The concept of taxonomic predictivity proposed by Jansky and co-workers to target species for trait evaluation would be aided by molecular characterization within selected taxa. High density genotyping systems, which target known chromosomal locations, such as the Illumina Infinium SNP genotyping platform being developed by the SolCAP initiative or markers that are more cost-effective for whole-genome profiling, such as the potato DArT array, will contribute to germplasm characterization and selection of new genetic resources for pre-breeding activities.

These genome-wide markers will also be useful for selection against donor genome content during a backcrossing program. Screening for variation in candidate genes will be greatly facilitated by the completion of the potato genome sequence, next-generation sequencing technologies, and novel approaches to variant scanning, such as the use of HRM. The introgression of beneficial traits from wild potato species has several biological and practical hurdles. Wild species offer a new source of variation that can be selected within a breeding program; however, this variation is often introgressed with reproductive or phenotypic characteristics that inhibit selection of individuals as progenitors of subsequent breeding cycles.

A better understanding of the genetic mechanisms underpinning reproductive barriers and adaptation traits, such as photoperiod response, plant architecture, and tuber morphology may also provide a source of markers that can be used to screen early backcross generations. Conventional, marker, and transgenic approaches to enhance meiotic recombination should also be developed to enhance the probability of identifying individuals with smaller fragments of the donor genome. There are several opportunities to apply cisgenic technology to access beneficial diversity in wild species and these will be discussed in Section 3.5.

Parent Development with DNA Markers

Parental line breeding in hybrid potato cultivar development generally represents a smaller component of potato breeding programs relative to the selection of potential cultivars. For qualitatively inherited traits, such as the phenotype imparted by a dominant resistance allele, the benefits of developing parents with two or more copies of a favorable allele are evident by the predicted increase in proportion of progeny with the desired allele. Despite these advantages, the conventional development of parents with duplex genotypes is very difficult since it requires progeny testing before the parental genotype is known.

Once a parent is identified with two or more copies of a resistance allele, there is a tendency among potato breeders to re-use these parents, even when these parents produce progeny which are discarded due to undesirable characteristics. Such markers should promote the notion of forward selection of parents within potato breeding programs. The use of markers will also allow crosses to be designed that pyramid resistance genes for a single pathogen, thereby improving the chances of selecting cultivars with more durable resistance.

Quantitative genetics theory predicts that cultivars with the best phenotypic performance will not necessarily be the best parents for subsequent generations in a heterozygous autotetraploid crop. Genomics technology may help clarify the distinction between parental line development and cultivar development by determining the molecular basis of heterosis and the relative roles of additive, dominant, and epistatic genetic controls. As gene action in autotetraploid potato is better understood, markers will likely play an even greater role in parental development in the future. The greatest impact of markers will be for traits predominantly under the control of additive genetic variation, since they will contribute the most to heritable differences upon which selection may act. One potential application of MAS to parent development is the replacement or complementation of pedigree-based relationship assessments with one derived from genome-wide marker information. This information has been used in association genetics studies to generate mixed-model equations to identify markers linked to quantitative traits. Best linear unbiased predictors of breeding value that rely on pedigree-based calculation of a relationship matrix are routinely utilized in animal breeding to select superior parents. This methodology has also been shown to have great potential for identifying superior parents within a potato breeding program. As evidenced by the unexpected results from analysis of the origin of long-day adapted Neo-tuberosum germplasm, early pedigrees from progenitor clones may be very inaccurate. It is also known that the covariance among relatives in autotetraploids is much more complicated than in diploids, so it is conceivable that marker-based relationship matrices may be superior for breeding value calculations than those derived from pedigrees.

Early-generation Selection with DNA Markers

The application of MAS in early generations of a potato breeding program requires a careful cost—benefit analysis. On one hand, selection in the first field generation is quite ineffective, especially for traits with mid to low heritability and marker-based selection may be more effective than visual selection. On the other hand, the selection intensity is very high in the early generations, which would result in a considerable expense for marker genotyping clones that would be subsequently discarded for a critical trait. Therefore, it would be more practical to initiate marker genotyping after the first field generation when 90–95% of the progeny are discarded, as suggested by Ottoman et al..

At this stage of the program, the use of markers linked to highly heritable traits, such as major resistance genes, would be recommended. Markers linked to QTLs controlling a relatively small portion of the phenotypic variation of a trait should not be utilized in the early generations. Traits selected by breeders in the early generations have generally not received the research attention devoted to more economically important traits. A better understanding of the genetic mechanisms regulating traits which form the basis of a "discard" decision in the first or second field generation may lead to the development of a sufficient number of markers to warrant implementation of MAS sooner. Data are not typically collected from individual clones in the first field generation; therefore, these marker-trait associations will likely have to be identified in genetic studies which are separate from the breeding program.

Combined with the selection of parents with higher breeding values for adaptation traits, MAS could facilitate the increase in proportion of desirable progeny to be advanced to later generations. This should result in better use of field resources and increased population sizes to more effectively select difficult traits. Advances in DNA extraction technology will also contribute to increasing the feasibility of MAS in early generations. The utilization of automated tissue grinding equipment with the capacity to simultaneously extract DNA from samples in 96-well plates has greatly increased the throughput capacity of our potato molecular breeding laboratory. Further capacity may be obtained using a liquid-handling robotics workstation.

Advanced-generation Selection with DNA Markers

At the advanced stages of a breeding program, substantially more phenotypic data are available for consideration compared to early generations. It is logical then that marker information will be combined with trait data and perhaps pedigree or marker derived relationship data to identify selections to be promoted and/or released to industry. Markers may also play a role in identifying clones which will undergo a particular evaluation. These decisions are often based on a limited amount of data from the parents since these trials can be very expensive or difficult to undertake. The use of the *H1*—linked HRM assay has identified parents in the Agriculture and Agri-Food Canada breeding program that putatively contain the *H1* allele for resistance to golden nematode pathotype *Ro1*. Progeny containing this *H1*—linked marker are entered into resistance evaluation trials, regardless of whether the phenotype of the parent is known. Each breeding program will have its own way of incorporating marker data into selection decisions. Marker data would fit nicely into selection indices that incorporate information from multiple sources. The molecular eigen selection index method proposed by Cerón-Rojas et al. has some practical advantages, since it does not require economic weights of the traits under selection. For any marker-based selection approach, it will be important to take into account the proportion of phenotypic variation explained by a marker, the number of markers evaluated per trait, and the existing phenotypic performance data for the clone.

Enhanced informatics capacity would allow breeders to have access to databases with marker, phenotype, and pedigree information. Such an information/data analysis system would facilitate "real-time" assessment of individual markers in the context of genetic relationships, environments and other data to develop a value that could be used to rank clones each year prior to planting. At this time, the major factor limiting applications of MAS within breeding programs, especially at the advanced generation stage, is that there are relatively few validated, user-friendly markers linked to QTLs. Association genetics studies involving advanced breeding lines have contributed greatly to the tool-box of available markers for quality and disease resistance. The Potato Genome Sequencing Consortium will enhance our knowledge of the potato genome and initiatives like SolCAP will further bridge the gap between breeders and biotechnologists.

TRANSGENIC BREEDING

Classical breeding is most often thought of as the crossing of two individuals of the same species or closely related species, with the resulting progeny having a mixture of genes from both parents and tens of thousands of genes involved. Classical breeding is often inefficient and imprecise as the breeder has very little control over which genes will be transferred to the next generation or how to track them; although the use of markers will improve precision. Multiple years of evaluation are often needed to select a few candidate lines for further analysis from the pool of progeny generated. Clonal propagation and a tetraploid genome add another layer of complexity to the process.

With transgenic breeding, a specific gene or genes from a related or widely divergent species can be moved into potato, often into the cultivar of choice without backcrossing. As the potato industry has been slow to adopt new germplasm, the ability to add the desired trait to a selected cultivar is an added benefit. However, the newly created transgenic line must undergo years of testing and regulatory hurdles to show that the introduced genes carry no deleterious effects to either the plant itself or the consumers and users of the plant. In this section, we will review the use of transgenic breeding in plants, and more specifically, in potatoes.

In potato, *Agrobacterium*-mediated transformation is the most common method for transferring genes in transgenic breeding. *Agrobacterium tumefaciens*, a common soil bacterium, transforms many dicot plants, including potato, with relative ease. Select bacterial genes are removed and the genes of interest are inserted in the bacterial plasmid. The bacterium orchestrates the random insertion and integration of the selected DNA into the genome of the plant. Expression of the gene or genes is dependent on many factors including location in the genome and type of promoter used to drive the gene.

A screenable or selectable marker gene is often linked to the gene of interest to be able to determine if the plant is carrying the intended DNA. The most common selectable markers are usually antibiotic or herbicide resistance and bacterial in origin. For potato, the *nptII* gene is frequently used. *nptII* confers antibiotic resistance to the plant and allows the expressing plant to survive on kanamycin.

Other genes and gene components are often bacterial (*Bt*) or viral in origin as are the promoters. The use of widely divergent genes has allowed many unique traits to be moved into plants to improve quality, provide resistance, and provide stress tolerance. There are a wide variety of traits in potato that are currently being targeted using transgenic breeding. A partial list of traits includes: starch; late blight resistance; virus resistance; carbohydrate metabolism; bacterial resistance; insect tolerance; molecular farming; bioplastics; drought tolerance; salt tolerance; improved nutrition; and functional foods. One of the most notable uses of transgenic breeding in potato was Monsanto's NewLeaf by NatureMark™ potatoes.

Monsanto released several varieties of the NewLeaf potato with CPB resistance alone or in combination with either Potato Virus Y or Potato Leaf Roll Virus resistance. CPB resistance was conferred by a bacterial gene (*Bt*) and the first line was released in 1995, followed by lines with virus resistance (conferred by viral coat protein genes) over the next several years. In 2000, McCain Foods announced that it would no longer accept NewLeaf potatoes due to concerns over consumer acceptance from the quick food industry.

NatureMark was dissolved in 2001 and the potato project discontinued due an 80% decrease in sales of the NewLeaf varieties in 2000. Monsanto continues to license both the insect and viral resistance genes to other organizations outside of the US and Canada. BASF Plant Science Company currently has two transgenic potato varieties in development. "Amflora", an industrial modified starch variety, is under development for the past 12 years and awaiting approval since 2007. "Fortuna", a late blight resistant variety, is several years away from approval at the time of publication. The "Amflora" application has been reviewed and the variety has been deemed safe by the European Food Safety Authority.

However, continual delays in approval from the European Commission (EC) have led the company to file a lawsuit against the EC that have resulted in the suspension of further research by the company to develop additional late blight resistant varieties until issues with the approval process have been resolved. Consumer acceptance of transgenic crops varies from country to country and from crop to crop. Many North American consumers are comfortable with the use of transgenic plants, also known as genetically modified organisms, as a food ingredient. However, there is uncertainty regarding consumer acceptance of GMOs when consumed as a whole food such as potato, banana, and papaya; especially when the consumer sees no immediate personal benefit.

Rather than risk losing any market share, the quick food industry has chosen to avoid using GMOs and transgenic potatoes have disappeared from the market place. Additionally, consumer response to GMOs in Europe and other parts of the world has been less favorable and has led to the exploration and development of "greener" methods of creating transgenics. Greener methods are expected to have greater consumer acceptance by using genes from more closely related species of plants or from the same species and, in some jurisdictions, reduce the regulatory hurdles to approval. The word "transgenic" indicates a transfer of genes across species, such as a bacterial gene introduced into potato. New terminology, "cisgenic", has arisen recently to indicate that the movement of genes is occurring within the same species or plant family. There are variations on this theme, but the basic elements are that the genes being transferred into the plant come only from plant sources.

There are now several reports in which *Agrobacterium*mediated transformation has been used to create cisgenic plants. Other ways to make the genetic transformation process more acceptable to the consumer is to avoid the use of a selectable marker, such as antibiotic resistance; to remove the selectable marker before the plant goes to the consumer; or to use a selectable marker that is plant based. Transgenic or cisgenic breeding methods have the potential to be another tool for use in potato improvement. However, there are limitations and benefits to both classical and transgenic breeding. Research on transgenic potatoes continues with a wide variety of biotic and abiotic stress resistance traits being developed even though the current market is limited. The ultimate goals of the breeder will determine which method will best help to achieve these goals.

4

Climate Change: Impact and Opportunities for Potatoes

WHAT ARE THE OPTIONS FOR AGRICULTURE?

Given that certain effects of climate change are already being felt, climate change must be addressed with a combination of effective climate protection measures, on the one hand, and adaptation measures, on the other hand. Basic climate protection measures and binding objectives were established in the Kyoto Protocol to the United Nations Framework Convention on Climate Change. In the agricultural sector, several organisations – including the Food and Agriculture Organisation (FAO) and the international agricultural research centres of the Consultative Group on International Agricultural Research – have elaborated measures to support the most severely affected countries.

ADAPTATION

Climate change underlines the importance of sustainable agricultural production systems. The more stable an agricultural ecosystem, the better it can cope with the additional stress factors induced by climate change. Sustainable resource use and sustainable land management are efficient ways to strengthen an agricultural system. Each of the individual measures to enhance sustainability has several positive effects. Mulching, for example, can raise the content of organic matter in the soil. This, in turn, increases the soil's water holding capacity, thus reducing runoff and erosion and making more water available to plants.

Sustainable water use can also be achieved through improved irrigation systems, one example of which is drip-irrigation. Such systems need not be expensive. In some developing countries such as, for example, India, low-cost micro-irrigation systems are produced locally and at a relatively moderate price. Mixing varieties and intercropping (growing multiple crops on one plot) lead to more biodiversity within a plot, thereby also reducing this plot's vulnerability to different weather impacts. In many cases, research can learn from the knowledge of farmers in marginalised regions. Among other things, farmers adapt the sowing time to changes in the onset of the rainy season, or they react to shorter growing seasons by adapting their cultivation methods. These measures are not always sufficient. Further steps must be taken. High importance is ascribed above all to developing and breeding heat and drought resistant varieties. Furthermore, there is a need for strengthening rural institutions, and for access to micro-credits, micro-insurance, and climate information. All these measures require funding and support by national and international organisations. Particularly the most heavily affected poor rural populations are in urgent need of such support.

MITIGATION

Agriculture has the potential of reducing greenhouse gas emissions without making large financial investments. Often, these measures have additional positive effects. The most crucial issue is reducing forest clearances for agricultural purposes. Agroforestry systems can be a viable alternative. Moreover, zero tillage can raise soil organic matter content, at the same time preventing erosion and binding CO2 in the soil. New rice varieties that require less water also reduce methane emissions, and reduced nitrate loss during fertilising also diminishes emission of N2O.

A NEED FOR CONCRETE ACTION

VULNERABILITY ASSESSMENT

International and local institutions, as well as governments, must be aware of the impact of climate change. They must include these impacts in their planning and devise measures to counteract them. However, this is difficult: *"...at present, there is virtually no experience of how to manage this adaptation process. There are scarcely any systematic risk assessments predicting who will be affected and when..."* In a first step, vulnerability to climate change must be assessed. Only once it is known how heavily the various areas, crops, population groups, etc. will be affected by the impacts of climate change, is it possible to define measures.

Several tools have been developed to assess the vulnerability of individual projects. Two of them shall briefly be presented below. CRiSTAL is a computer-based decision-support tool that helps to reveal risks related to climate change in project planning.

In a second step, the tool suggests options for adaptation. The objective is to illustrate the interlinkages between climate, livelihood, and project activities. ADAPT, developed by the World Bank, is a computer-based tool to screen development projects for climate risks. The programme uses various climate change scenarios, linking them to planned project activities in order to determine the project's vulnerability to failure due to climate change. The tool is intended for use by development cooperation practitioners who are planning new projects. The International Potato Center (CIP) has developed a simulation model that succeeds in simulating and visualising different climate scenarios and their influence on potato production.

The model is based on a geographic information system (GIS) in which users can enter the variables for precipitation, temperature, soil, and different potato varieties. The output consists of estimates regarding the future development of potato production, including yield estimates, for the given region. The possibility of downscaling and interpolating available climate data has greatly enhanced the precision of projections and makes the model applicable to much smaller areas. Farmers and researchers can therefore use it as a support tool for making decisions with regard to potato growing. Simulation models are also used in breeding new varieties in order to simulate the impacts of new heat and drought resistant cultivars under certain climatic conditions.

THREE SCENARIOS FOR POTATO FARMING

Depending on a region's vulnerability, different measures need to be taken. Potato producers, for example, face three options for reacting to the changed conditions.

Abandonment of Production

Should climate change lead to a drastic deterioration in growing conditions, farmers in certain regions will have to consider abandoning potato production. The Sikasso area in southern Mali can serve as an example. In recent years, the government and international donors supported strong promotion of potato production and marketing in the area. Due to climate change, these efforts are now put into question, and project partners are thinking of stopping potato production altogether as an ultimate adaptation measure.

Expansion into New Areas

In other areas, potato cultivation is only being made possible or can be intensified due to climate change. In some regions in China, for example, potatoes are expected to become a possible winter crop. However, when introducing potatoes as an entirely new crop, it is crucial to consider possible negative consequences.

Adaptation of Production

In many areas adaptation measures will be necessary in order to maintain production or reduce yield losses. Several possible adaptation measures and considerations regarding potato production are presented in the following section.

GROWING IMPORTANCE OF WEATHER FORECASTS

Information on climate change and its impact are of great value to farmers. For example, early warning systems can alert the population about imminent storms and enable them to take protection measures in time. Medium range weather forecasts can be particularly useful for planning crop schedules. In many places, farmers rely heavily on traditional weather forecasting methods.

Climate change is increasingly causing these methods to lose their validity, and the weather is becoming more difficult to predict. As a result, scientific weather forecasts are gaining importance.

However, many farmers are sceptical towards scientifically produced forecasts despite the fact that their traditional forecasts are becoming less and less reliable. A study reports farmers in the Andes to have claimed that forecasts produced in distant weather stations were of no use to their region. As a rule, they relied much more on forecasts by local experts than on those produced by professional weather services.

For planning their crop schedule they usually looked to the most successful producers in the area, whose short- and long-range weather forecasts were based on observation of the stars and other natural phenomena. In order for scientific weather forecasts to benefit the farmers despite their scepticism, local experts must be convinced to cooperate and help professional weather services to reach farmers better by providing locally adapted solutions.

BREEDING NEW VARIETIES – A KEY TO SUCCESS

Both researchers and farmers see an urgent need for breeding new potato varieties that are better adapted to the changing climatic conditions. Particular efforts in this respect are made at the International Potato Center (CIP), where the main focus is on breeding short-season varieties.

Due to their shorter vegetation periods, these varieties make it easier for farmers to avoid unfavourable conditions such as hot or dry periods and react to modified rainfall patterns. Climate change confronts breeders with an enormous challenge.

Normally, they work to improve one specific characteristic of a crop, whereas climate change has created a need for adapting several characteristics at once. Therefore, the CIP is evaluating the genetic resources of its potato collection, and also of new varieties still in the process of being bred, with regard to their tolerance of water and temperature stress.

The CIP can draw on the world's largest genetic reservoir of potato varieties: its genebank contains 5000 distinct types of cultivated potatoes, as well as more than 2000 wild relatives of the potato belonging to around 140 wild species. The goal is to identify the desired key characteristics and genes that determine tolerance of abiotic stress by means of various selection methods and marker systems.

EARLY-MATURING POTATOES FOR SHORTER RAINY SEASONS

The Fundación para Promoción e investigación de productos andinos in Bolivia, together with the CIP, is in the process of launching a new potato breeding project in order to be able to react to climatic changes. In recent years, some areas in Bolivia have been experiencing changes in the onset of the rainy season. Previously, farmers would start planting their potatoes in October, at the onset of the rain. Today, the rainy season often begins as late as December, whereas it still ends in March, as it always has, confronting farmers with drastically shorter growing periods.

Traditional potato varieties are not adapted to these short rainy seasons and thus produce low yields. Further progression of climate change can therefore threaten the food security of large parts of the population. Until now, many potato farmers have been growing several varieties on one plot in an attempt to minimise the risk of yield loss due to diseases, pests, and adverse weather. Jorge Rojas of PROINPA doubts that this will be sufficient as an adaptation measure in future. According to his observations, climate change appears to take a particularly radical form in the Andes, not leaving farmers enough time for slow adjustment. They need assistance and support, particularly with regard to breeding new varieties. PROINPA has launched a project to breed potato varieties that are better adapted to the short rainy season and do not produce lower yields due to the new seasonal regime. The project is designed to involve farmers very closely by testing new varieties on their fields, under real conditions, as early as possible.

WILD RELATIVES – AN IMPORTANT GENE POOL

Success in breeding better adapted varieties depends on collection, preservation and dissemination of the relevant genetic material. Wild relatives play an important part in breeding abiotic stress tolerant varieties. Their great genetic variability makes them important genetic resources for producing the desired characteristics. Until now, wild relatives have rarely been used for breeding purposes because they frequently have many undesirable characteristics aside from the desired ones. Breeding with wild relatives still needs time to be further developed, since it requires sound experience. Climate change and other factors that additionally increase pressure on ecosystems are threatening the existence of many wild relatives. The establishment and maintenance of genebanks is intended to curb the loss of this diversity in varieties. The genetic resources of the potato are preserved in the form of true potato seeds, vegetative tubers, and in vitro seedlings. With regard to potatoes, in particular, conservation under in situ conditions is considered an important strategy to preserve the genetic resources. In situ conservation involves exposing the varieties in question to natural conditions in the field. This initiates a process of selection and, perhaps, even adaptation to the changing climatic environment. Experts have devised strategies for improving conservation and use of crop genetic resources. To date there are around 1400 genebanks worldwide, containing approximately 6 million samples of crop genetic resources. Three quarters of them are located at CGIAR centres. In future, genetic resources for such collections need to be selected more according to the newly required crop characteristics, such as, for example, drought tolerance. In addition, collections and particularly their diversity must be extended to include more wild species and land races. In view of the fact that it takes 12–20 years to breed a new variety, researchers have not been able to answer the question of how well the existing genetic resources are suited for meeting the challenge of climate change.

CLIMATE CHANGE AND POTATOES

CLIMATE RISKS TO PRODUCTION

Photosynthesis is the driving force in production and governed by incoming levels of solar radiation. However, the production potential is also influenced by temperature and water availability, technology, fertiliser and crop losses. As well as the direct impacts of climate change (i.e. rainfall and temperature), potatoes will also be particularly sensitive to various indirect effects, including changes in the agricultural potential of soils, as this effects the availability of water to plants and impinges on other land management practices (e.g. trafficability for seed bed preparation, spraying, harvesting). Increases in atmospheric CO_2 concentration will also impact on crop growth by increasing the resource efficiencies for radiation, water and nitrogen. As a consequence, for most crops grown in northern Europe including potatoes, future warmer temperatures and elevated CO_2 levels are expected to result in more favourable growing conditions, although of course there will also be some negative consequences. The latest climate projections for the UK suggest drier summers with higher temperatures and reduced rainfall. But note the uncertainty around these 'mean' projections.

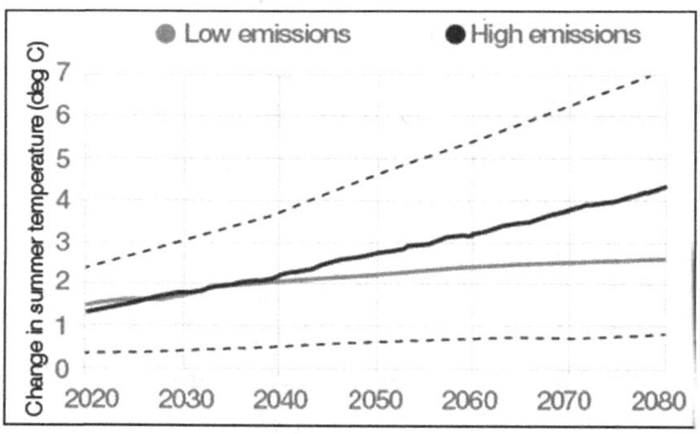

Fig. Projected changes in mean summer temperature (2020s to 2080s), for the low and high emissions for a farm site in Suffolk. Dotted lines represent extreme probabilities.

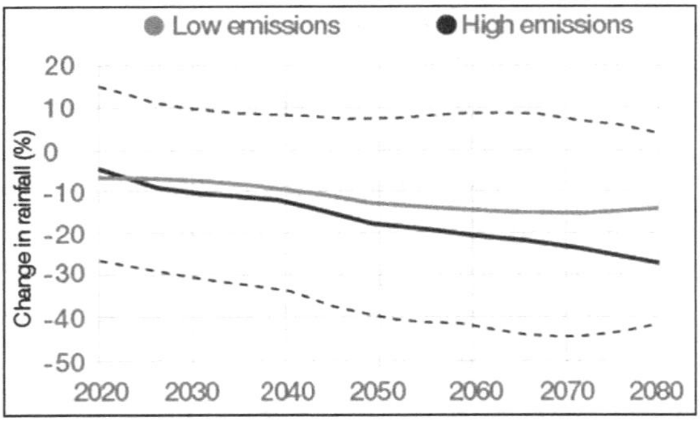

Fig. Projected changes in mean summer rainfall (2020s to 2080s), for the low and high emissions for a farm site in Suffolk. Dotted lines represent extreme probabilities.

In general, at higher latitudes a rise in temperature tends to increase the developmental rate of the crop and extends the length of the growing season, resulting in a positive impact on crop production. On the other hand, reduced summer rainfall is likely to increase soil moisture deficits reducing yield under rain-fed regimes and increasing the need for supplemental irrigation.

FUTURE POTATO YIELDS

To assess yield impacts, a potato crop growth model was used. This simulates the daily growth and development using information on climate, soil, management and cultivar. The model was first calibrated using field data from three farms (in Norfolk, Suffolk and Lincolnshire) and an experimental farm (CUF) to check that the modelled yields accurately reflected those observed on-farm. Future potato yields were then simulated for each farm for selected emissions scenarios in the 2050s.

The projections showed only minor increases in potato yield, mainly in response to the increased radiation levels and higher temperatures. Farm production levels were found to be limiting due to current water and fertiliser management practices. But when future production assumed an optimal irrigation and fertiliser regime then future average yields on-farm increased by +13 to 16%.

- For growers, climate change presents an opportunity to increase potato yield, but getting the right amount of both water and fertiliser applications will be key. To achieve this, growers will need better information to manage climate uncertainty and match crop demands to environmental conditions.

FUTURE IRRIGATION NEEDS

Changes in climate mean supplemental irrigation is likely to become more widespread, particularly on varieties where scab control for skin finish is important. For a pre-pack Maris Piper, the combined effects of reduced rainfall and increased crop evapotranspiration led to a significant increase in future irrigation need, ranging from +14 to 30%.

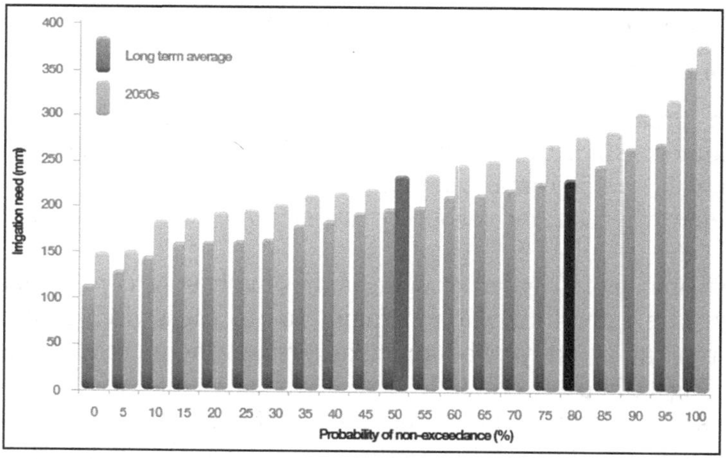

Fig. Projected future changes in annual irrigation need (mm) for pre-pack Maris Piper.

Clearly, these increases could have major cost implications – both on production, water resources and existing infrastructure. Current irrigation system capacities are typically designed for a 'design' dry year (i.e. 80% probability of nonexceedance). A future 'average' year could be much drier than a current 'design' dry year.

- Without adaptation, existing irrigation schemes could have insufficient capacity to meet future irrigation needs in approximately 50% of years, with serious consequences for potato yield, quality and farm income.

IMPACTS ON LAND SUITABILITY AND SOIL MANAGEMENT

Land Suitability Classes

Four classes are frequently used by soil scientists and growers to assess land suitability for potato production - well, moderate, marginal and unsuited. *Well suited* has a high and sustainable production potential from year to year with adequate opportunity to establish the crop in average years at or near the optimum sowing time. Harvesting is rarely restricted by poor ground conditions. *Moderately suited* is where potential production is variable from year to year due to a shortage of soil water to sustain full growth, or poor conditions at crop establishment.

Marginally suited has a potential production that is variable from year to year with considerable risks, high costs, or difficulties in maintaining continuity of output, due to climate interacting with soil properties, disease or pests. For potatoes, the criteria for *unsuited land* relate mainly to climate, gradient and stoniness.

Three factors are then used to distinguish between unsuited and suitable land; namely, potential soil moisture deficit (which reflects agroclimate conditions), accumulated temperature and slope.

Future Land Suitability

In many areas, rainfed production will become increasingly risky. However, with supplemental irrigation, around 85% of the total arable land in central and eastern England would remain suitable for production, although most of this is in catchments where water resources are already over-licensed and/or over-abstracted.

The expansion of irrigated cropping is thus likely to be constrained by water availability. The growth in water demand due to the switch from rainfed to irrigated cropping is likely to be much larger than the increase in need of the currently irrigated potato crops.

- In some regions rainfed production will become limiting – our results suggest that by the 2050s, the area of land that is currently well or moderately suited for rainfed production will decline by 74% and 95% under the "most likely" climate projections for the low and high emissions scenario, due to increased droughtiness.

Grower Implications - Variety Choice

The full impacts of climate change on land suitability are clearly more complex than the basic maps shown here, but the results agree well with qualitative judgement. For example, based on PCL data, the cost of production would not be covered by the crop value when maincrop yields fall below 30 t/ha. This would set a limit for rainfed production in the dry regions and those having coarse and easily drained soils.

A restricted water supply during crucial times can also damage tuber quality to the extent that certain varieties would be rejected by the market. This would force growers to shift to varieties less susceptible to scab or towards the processing market. Any reduction in irrigation availability or reduction in rainfall would severely affect the profitability of crops such as Maris Piper and Maris Peer, where skin finish is crucial for packing.

Determinate crops, i.e. those that only produce a limited leaf area and have short periods of active root growth and some less-determinate varieties are very sensitive to water restrictions during mid-late canopy expansion. Current widely grown examples include Estima, Lady Rosetta and Saturna. Absence of rain or irrigation during these periods can cause premature senescence with a large yield loss.

For this reason, rainfed or limited irrigation production with these varieties under these climate change projections is likely to be reduced for risk of crop failure. However, the yield response to irrigation of many of these varieties is large so they will continue to be grown where irrigation is less limited. Production in rainfed areas is thus likely to change to varieties that are able to either (a) survive early drought periods, so that they can use rainfall later in the season (e.g. Maris Piper, King Edward, Markies, Russet Burbank or Rooster) or (b) partition dry matter towards tuber production during periods of drought rather than canopy production which makes them more efficient in producing yield per unit of water.

ADAPTATION – COPING WITH GREATER CLIMATE UNCERTAINTY

Crop Management

As with all climate change impact assessments, growers need to be wary of the results and interpret them with caution. It is nevertheless important to consider the range of adaptation options and responses available, then weigh up the risks and plan accordingly.

The outputs from our crop modelling assumed unchanged farm practices in the future, but in reality there will of course be some degree of autonomous adaptation even if not planned adaptation. For potatoes, this could include earlier planting and harvest dates, changing to better adapted varieties, less dependence on soils with low water holding capacities, crop movement to regions with more suitable agroclimate conditions and water availability, and probably the uptake of GM technology.

Changes in Soil Management

Climate change is likely to lead to the dates of the last spring frosts becoming earlier and autumn frosts becoming rarer and/or later, thereby extending the growing season. Planting could therefore take place earlier as the thermal environment experienced by crop canopies would be more favourable. However, soils would still be at field capacity, leading to the same problems in workability that growers currently experience during March and April in many regions of the country. Reduced rainfall and higher temperatures will result in a depletion of organic matter, increasing the risk of structural damage to sensitive soils. Harvesting windows would become longer, thereby reducing the risk of adverse soil conditions causing harvesting problems or crop damage.

Land Management

The land suitability maps clearly show that growing rainfed potatoes in England and Wales will become increasingly risky as a result of climate change, and limited to a few favourable areas. In contrast, with irrigation the land suitability hardly changes and most of the current rainfed potato production could remain in its present location, if it could be irrigated.

Although only around 1% of water abstraction in England and Wales is used for irrigated agriculture, there is limited prospect of the industry obtaining significant additional licensed quantities for the summer months in the face of competing demands. However, many existing licences are unused or underused, so water transfers or abstraction licence trading between farms may be an option, though there are environmental arguments against re-activating "sleeper" licences in stressed catchments.

Water Management

Licences are still available for high flow abstraction in most catchments, and recent years have seen a significant increase in winter filled onfarm reservoirs for irrigation use in the summer. Though expensive, this gives the grower greater security of supply. It seems likely that this will become the preferred irrigation water source for potatoes and other high value vegetables in the south and east of England.

But there are still relatively few examples of adaptation to climate change in the UK agricultural sector, and that many apparent adaptive actions have actually been in response to legislative or other pressures, rather than purposeful adaptations to perceived climate change per se, but nevertheless they may still be useful climate change adaptations. The current growth in on-farm reservoirs would appear to fall into this category.

Investing in Water Efficiency

Once irrigation water is assured but expensive, it will become sensible to invest more heavily in water efficiency measures; better application methods, including drip and precision irrigation, and scientific scheduling methods will become standard.

Earlier planting and harvesting would reduce water use per unit area, but with some varieties growers might prefer to use the longer growing season to increase yield.

There has been a steady increase in average potato yields over the last 40 years; with national consumption roughly constant this has led to a gradually reducing area planted; whether this trend can be intensified and how far it could counteract increasing water demand is not yet clear.

Others have suggested that irrigated production might move north and west as an adaptation to climate change.Given that most of the current locations remain suited to irrigated production this may be a slow process. Many growers have sizeable investment in fixed assets, and may therefore prefer to remain at their present locations renting land from neighbours with unused or partially used licences as a preferred way forward.

Short Term Coping or Long Term Planning?

Greater uncertainty in seasonal weather patterns means growers need to consider short-term coping strategies as well as longer-term strategic developments to reduce their vulnerability to changing water availability.How they respond will depend to a large extent on their perception of risk and the opportunities that climate change presents to their business. Farmers generally have two options; either to reduce their water needs or try to secure additional water supplies.

Options to reduce on-farm water needs include investing in improved irrigation technology and equipment to increase application uniformity and efficiency, using weather forecasting to increase the effective use of rainfall, encouraging deeper rooting of crops, introducing lower water use or drought tolerant crop varieties, decreasing the overall irrigated area, or modifying soil structure to improve soil moisture retention.

Options to obtain more water include purchasing land with water, obtaining additional licensed capacity and building on-farm storage reservoirs (either individually or shared with neighbouring farms), installing rainwater harvesting equipment, re-using waste water from farm buildings, or switching water supplies to public mains where feasible.

- Many of these potential adaptations are already 'no regret', in that they already make sense by solving existing water resource issues, which then contribute to a farms' future adaptability.

Dealing with Uncertainty

It is important to remember that the potential impacts of climate change are based on long term projections – they are not predictions of what 'will' happen but rather projections of what 'might' happen. There is a large degree of inherent 'uncertainty'. Figure shows, for example, the projected changes in annual rainfall at Cambridge together with the uncertainty.

The important message here is that the uncertainty is greater than the projected future change – in August, for example, it is projected that rainfall might decline by around 20%, but it could range from -50% to +40%. This makes adaptation planning very difficult, particularly where large investments are being considered.

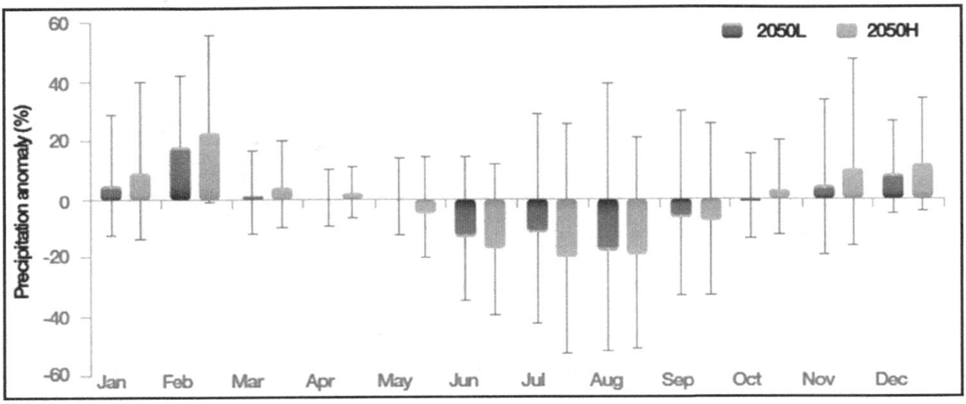

Fig. Projected change (%) in mean monthly rainfall at Cambridge from the current baseline in the 2050s.

THE EFFECT OF CLIMATE CHANGE ON GLOBAL POTATO PRODUCTION

It is likely that the currently observed trend of global warming, which has been 0.6 C 0.2 since 1900, will continue and that the average global temperature will increase by between 1.4 and 5.8 C over the period 1990 to ?lo0. The impact of this type of clim4te change will probably lead to a decrease in crop producbi@, but with important differences between regions.

The effects of climate change on crop production can be complex. Depending on the temperature regime and the crop, high temperatures can lead to low yields due to increased development rates and higher respiration.

However, a short growth cycle can also be beneficial, e.g., to escape drought or frost, and the use of late-maturing cultivars could offset the effect of high development rates.

In environments where low temperatures now limit production, global warming could lead to a beneficial lengthening of the growing season and temperatures close to optimal for assimilation. Moreover, global warming is related to the increase of atmospheric CO, concentration, which is likely to increase crop yields, particularly when water limits crop production. Potato is grown in many different environments, but it is best adapted to temperate climates. At high temperatures tuberization diminishes. Potato is also frost sensitive, and severe damage may occur when temperature drops below 0 C mmans et al. 2003).

Various authors have used simulation models to study the effect of global warming on potato production. Higher temperatures are predicted to increase potato yields in England and Wales, Scotland and Finland, primarily because of a longer growing season. However, an overall yield decrease was predicted for the USA. *AU* these studies on potato were conducted for small regions at high latitudes and their results are difficult to extrapolate to other regions.

In this paper, some of the possible effects of climate change on potato production are studied at the global level. A simulation model was med to calculate potential potato yield for the current climate and for projected future climates in 2010-2039 and 2040-2069, using seven climate scenarios from five different climate models.

The goal of the paper is to identify regions where there is likely to be a strong decline in productivity due to the increase in temperature, and to determine the extent to which heat-tolerant potato cultivars would be useful to mitigate the effect of climate change in those regions.

Only the effect of chwges in temperature and solar radiation was considered and hot the effect of changes in rainfall, of increased levels of atmbspheric CO,, or of increased ultraviolet radiation. The results h e presented in relation to the current global distribution of the potato crop.

CLIMATE DATA

Average monthly climate data for 1961-1990 (hereinafter referred to as "current climate"), and for climate change from that period to 2010-2039 and to 2040-2069 were used. For the current climate, data from New et al. were used; for the projected future climate, seven scenarios from five climate models were used.

The projected change in climate was superimposed on the current climate to create projected climate surfaces for 2010-2039 and 2040-2069 (further refereed to as 2010-39 and 2040-69, respectively). Daily temperature and radiation data were derived by linear interpolation between the monthly averages.

All datasets were resampled (statistically disaggregated by interpolation) to a 1 by 1 degree resolution, using IDRISI software (Clark Labs, Worcester, MA, USA). Global average temperatures for the current climate and for all scenarios were calculated for terrestrial cells only, without considering Antarctica, taking the size of each square degree grid cell into account (the size of the one square degree cells decreases with increasing latitude).

Table. Climate scenarios used in this study.

Scenario	Institute	Code	Forcing details	Original resolution in degrees	Solar radiation	Min. and max. temperature
I	Canadian Centre for Climate Modelling and Analysis	CGCM1	GS	3.75 x 3.75	Yes	Yes
II	Australian Commonwealth Scientific and Industrial Research Organisation	CSIRO-Mk2	GG	5.625 x 3.214	Yes	Yes
III	Australian Commonwealth Scientific and Industrial Research Organisation	CSIRO-Mk2	GS	5.625 x 3.214	Yes	Yes
IV	German Climate Research Centre	ECHAM4	GG	2.8125 x 2.8125	Yes	Yes
V	US Geophysical Fluid Dynamics Laboratory	GFDL-R15	GS	7.5 x 4.5	Yes	No
VI	UK Hadley Centre for Climate Prediction and Research	HadCM2	GG	3.75 x 2.75	No	Yes
VII	UK Hadley Centre for Climate Prediction and Research	HadCM2	GS	3.75 x 2.75	No	Yes

SIMULATION MODEL

The LINTUL simulation model as described by Stol et al. and Van Keulen and Stol was used to calculate potential yield. The model has a temperature-dependent development of the canopy. Biomass production is the product of the fraction green ground cover, incident solar radiation and radiation use efficiency. Stol et al. estimated tuber yield as a temperaturedependent percentage of the total biomass accumulated during the growing season. In this study, however, the relative allocation of biomass to tubers was calculated on a daily basis. Relative all@ cation to the tubers is initially *OO?*. After a thermal time threshold is reached, relative allocation starts increasing linearly with thermal time until the next threshold after which 1Wh of new bio mass goes to the tubers.

The values of these parameters were estimated *so* that harvest index of a mature crop is 80% under normal circumstances, as in the original model. This procedure avoids overestimating yield for a crop with prematurely killed foliage due to frost, or for a crop for which the end of the growing season is very warm, and hence has a lower b e s t i ndex than would be expected h m th e average temperature during the growing season. The absolute allocation of biomass to the tubers is also dependent on daily average temperature: it decreases above 15 C and becomes zero at an average temperature above 28 C. A heat-tolerant potaim cultivar was defined by shifting this curve two degrees.

In the adapted model, radiation use efficiency (RUE) was made dependent on daily average temperature, following Kooman and Kooman and Haverkort. RUE is highest (2.9 g MJ between 15 and 21 C md zero below 2 C and above 34 C, with intermediate value in between. Decrease of RUE at high temperatures is duel to increasing respiration. Radiation m2 day' was not considered,lto account for light saturation. For each grid cell, the model was run for 12 iplanting times (with planting at the first day of each month) and for five maturity classes of potato, representing different icultivars with early to late senescence.

This was repeated for lthe heat-tolerant potato. Maturity classes used were 1000,12b0, 1400, 1600, or 1800 Cd, expressed as the temperature sum between emergence and harvest, with a base tepperature of 2 C. The optimal planting time for a location @rid cell) was determined after the simulations, selecting the montNcultivar combination that led to the highest yield. Average temperature during the optimal planting time was calculated for each grid cell.

To distinguish between the effect of chanes in radiation and in temperature, the model was also run for the projected climate, while using radiation data for the current climate. Current and projected potential yield were compared for two cases: with and without adaptation. Adaptation is narrowly defined as changes in the month of planting or in the maturity class of the cultivar.

This is sometimes referred to as "autonomous adaptation" in the sense that these are inexpensive and can be carried out at the farm level. In the case of "without adaptation," potential yield for projected conditions is calculated for the culmination of cultivar and month of planting that gave the highest yield for the current climate. In the case of "with adaptation," the highest yield is taken from the 60 months of plantingkultivar combinations for the projected climate scenarios. Hence, in this latter case, the month of planting and cultivar type in a location can be different for current and future climates.

IMPACT OF CLIMATE CHANGE ON POTATO PRODUCTIVITY IN PUNJAB

GLOBAL climate change is an acknowledged fact and global greenhouse gas (GHG) emissions will continue for the next few decades. According to the Inter- Governmental Panel on Climate Change Report released in 2007, 11 of the previous 12 years ranked among the 12 warmest years in the instrumental record of global surface temperature. It is now reconfirmed that the global atmospheric concentrations of carbon dioxide (CO_2), methane and nitrous oxide have increased markedly as a result of human activities since 1750.

The increase in GHGs has resulted in warming of the climate system by 0.74°C between 1906 and 2005. The rate of warming has been much higher in recent decades. Although increase in atmospheric CO_2 has a fertilization effect on crops with C3 photosynthetic pathway and thus promotes their growth and productivity, on the other hand, it can reduce crop duration. Report of Working Group II of IPCC and a few other global studies indicate a probability of 10–40% loss in crop production in India with increase in temperature by 2080–2100. Potato is an important food crop of India.

We have made tremendous progress in potato production. The growth of potato in India has been phenomenal since 1950 and its per capita availability has increased from 4.37 to 21.52 kg in 2012. Climate change will have a profound effect on the potato growth story in India impacting every aspect of production and profitability, including productivity. The CO_2 concentration and assimilation are positively correlated and a 10% increase in tuber yield is estimated for every 100 ppm increase in CO_2 concentration. These positive effects are attributed to increased photosynthesis by 10–40%. Besides, elevated CO_2

concentration reduces evapotranspiration (ET) resulting in water saving to the extent of 12–14%.

However, increase in temperature has adverse effects on potato growth. High temperature is reported to reduce tuber numbers and size. Tuberization and gross photosynthetic rate are inhibited by moderately high temperatures, which ultimately affects the total biomass production and tuber yield. Increase in temperature and atmospheric CO_2 are both interlinked and occur simultaneously and the CO_2 enrichment does not appear to compensate for the detrimental effects of higher-temperatures on tuber yield. In India, potato is largely grown during winter season and is mainly confined to the Indo-Gangetic Plains. The autumn/winter planted crop in the northern plains of India comprising the states of Uttar Pradesh, West Bengal, Bihar, Punjab and Haryana contributes 90% of total potato production in the country. The availability of suitable growing period in India is likely to be impacted seriously by climate change and global warming. According to the IPCC Fourth Assessment Report (AR4), an increase in temperature ranging from 0.78°C during September–November to 1.17°C during December–February is expected under A1FI scenario by 2020 in South Asia. These changes are expected to aggravate and range from 1.71°C during June–August to 3.16°C during December– February, i.e. the main potato growing season in 2055. Thus, in 2020 the potato season is likely to be warmer by 0.78–1.18°C and in 2055, by 2.41–3.16°C under A1FI scenario.

There is an urgent need to study the impact of the likely changes in temperature and CO_2 on regional vulnerability of potato productivity in the future, in order to direct our research efforts to meet the challenges and devise adaptation strategies to minimize the impact of climate change. For this purpose, crop growth models are useful as they serve as an excellent tool to simulate crop growth and yield of annual and perennial crops under diverse situations.

These models are being used widely throughout the world for generating knowledge about the production potential of the crops under different climatic regions, for yield forecasting, nutrient and water needs of the crop, management decisions, impact analysis of climate change and adaptation strategies, selection of crops and varieties for optimum production, etc. Keeping the above points in view, a study was undertaken to assess the impact of climate change under A1FI scenario through simulation studies using WOFOST crop growth model on the productivity of three potato cultivars of different maturity groups in Punjab, which accounts for 83.6 thousand ha area under potato cultivation.

MATERIAL AND METHODS

Selection of Simulation Model

WOFOST model has been developed at Wageningen University, The Netherlands. It is a mechanistic model which simulates the growth of a crop based upon eco-physiological processes. The major processes are phenological development, CO_2 assimilation, transpiration, respiration, partitioning of assimilates to the various organs and dry matter formation. This model is widely used to assess the effect of climate change on the growth and yield of many crops, including wheat, rice, maize, potato, barley, soybean and sugar beet throughout the world. It is one of the three most widely used crop growth models for climate change studies. WOFOST has been calibrated for Indian potato cultivars using the time-course data on potato growth and development. Impact of climate change on potato productivity in the present study was therefore assessed using WOFOST model, which was run for 13 locations spread across Punjab.

SELECTION OF POTATO CULTIVARS

For simulation studies, three potato cultivars; one each from late, medium and early maturity group were selected. Together, these three cultivars account for about 90% of total potato acreage in Punjab. For climate change studies, the planting date was set as 15 October for all the cultivars and for all the locations since this is the present recommendation for planting of the main crop and the simulation was done for potential yields of potato cultivars for all the scenarios.

Table. List of meteorological stations located in Punjab for which simulation studies were carried out

Meteorological station	Longitude (°N)	Latitude (°E)
Amritsar	31.617	74.850
Bathinda	30.204	74.943
Faridkot	30.670	74.750
Fatehgarh	30.651	76.389
Firozpur	30.917	74.600
Gurdaspur	32.033	75.517
Hoshiarpur	31.530	75.920
Ludhiana	30.910	75.850
Mansa	29.980	75.380
Moga	30.480	75.100
Muktsar	30.283	74.305
Patiala	30.340	76.380
Sangrur	30.230	75.830

WEATHER DATA

India Meteorological Department (IMD) district normals of 1971–2000 of 13 districts of Punjab were used for baseline scenario in the present study. For calculating total solar radiation, Hargreaves–Samani equation was used which utilizes maximum and minimum temperature to estimate solar radiation and is reported to be best suited for Indian conditions. The weather data was then converted into WOFOST weather file format. The simulation studies were carried out for A1FI highemission scenario.

For generating the scenario for 2020 and 2055, projected changes in surface air temperature for sub-regions of Asia under SRES A1FI pathway, based on AR4 atmosphere–ocean general circulation models were added to the baseline data. Projected atmospheric CO_2 concentration based on the Bern-CC model for A1FI scenario was used for incorporating the effect of change in CO_2 concentration in the WOFOST model. The figures used in this study for atmospheric CO_2 concentration were 367 ppm, 415 ppm and 590 ppm.

INCORPORATION OF CO2 IMPACT IN THE MODEL

The impact was incorporated in the WOFOST model by making changes in light-use efficiency of single leaf and maximum leaf CO2 assimilation rate. Wolf *et al.* have reported that doubling of CO_2 from 355 to 710 ppm brings about 25–40% increase in yield in C3 plants and based on the literature reviews they have made changes to the initial angle and to the maximum CO_2 assimilation – light response curve parameters of WOFOST model for doubling CO_2 concentration from 355 to 710 ppm. Under experimental conditions with non-limiting supply of water and nutrients, where temperatures are kept near the optimum for crop growth, the yield increase for C3 crops and doubling of CO_2 has been estimated at 30% by various workers. Therefore, we have taken these figures as + 10% and + 55% for doubling CO_2 concentration for potato, assuming roughly linear relationship between CO_2 increase and growth processes. Accordingly, these parameters were changed for 2020 and 2055 for A1FI scenario. In the present study, the WOFOST model was run for 13 locations spread across Punjab, for all the three potato cultivars for baseline situation and for future climates, separately for temperature and CO_2 changes as well as for their combined effect, in order to assess the impact of the two parameters and their interaction with crop growth and productivity.

KRIGING

GIS maps of baseline productivity and change in productivity of different potato cultivars under future climate scenarios were generated using remote sensing and GIS software 'Geomatica'. An image of Punjab of 500 m pixel size was used for generating the GIS maps. A point layer of 13 districts of Punjab was created using their geographical locations.

The attribute data of each district containing productivity under different scenarios were geo-statistically interpolated using kriging technique to produce surface layers of the attributes. For kriging, spherical variogram was chosen for interpolation of the point data.

The modelling for per cent change in productivity was done in EASIPACE. GIS software was also used to estimate the area falling under different attribute classes.

GROWTH AND YIELD RESPONSE OF IRISH POTATO

Irish potato is said to have originated from the highland of Boloivia in South America. The spread of the crop outside its centre of origin was mainly by deliberate introduction. The crop moved out of South America to Spain in 1570; to England in 1585; then to Ireland by Spanish explorers from 1963. The crop was grown on a large scale in Ireland and became so popular such that it is not surprising that it acquired the misnomer of "Irish Potato". In development countries, Irish potato is ranked first in energy production per hectare per day, significantly above cassava and cereals.

It is a lover of cool climate and therefore requires a cool growing season with a moderate and well distributed rainfall of about 800mm during growing seasons with no prolonged dry weather. It could be grown under rain-fed condition or irrigated, but waterlogged areas are unsuitable. Temperatures higher than 270 C are unfavourable for the production of economic size tubers. Observations have shown that temperature ranges of 21°C – 26°C is required for sprouting of the tubers. Irish potato was introduced into Nigeria early in the 20th Century by European miners in Jos Plateau. Jos Plateau has high altitude and thus, cool climate, which is favourable for the development of the crop. Jos South Local Government Area (LGA) accounts for 25% of the total Irish potato produced in Nigeria.

The crop is efficient in converting land, labour, water and capital into a highly nutritious food. This is not surprising if for no other reason than it has a shorter growing cycle of about 95 days than most other tuber crops in the tropics. Irish potato is an important staple food as well as raw materials for industries. In order to meet the demand for industrial and human consumption, the yield per hectare needs to be improved. Such imminent improvement could be achieved through efficient management and monitoring of agro-climatic parameters, among others.

Several studies have been conducted towards improving the yield of Irish potato. For instance, Damkor based his studies on effect of stem thinning on yield and other vegetative characteristics; Ajala studied growth development and yield in four varieties of Irish potato; Szlachetha studied flowering behaviour of some varieties of Irish potato; Sale, Susnochi and Shimshi, Zaag and Burton, and Wolfe et al carried out a study on water stress and related analysis on different varieties of Irish potato. In Nigeria, Ifenkwe and Okonkwo conducted a study on determination of the most suitable time to plant Irish potato, while Ifenkwe did a comparison of flat lands and ridges for dry season planting on potato. Such studies have however, not been usually geared towards determining the relationship between climate and agricultural production, particularly considering Irish potato.

There is no doubt that analyzing the effect of climatic elements from the point of view on how they influence our agricultural activities is important in any meaningful plan for increase crop yield. Thus, microclimate studies have become imperative in our studies of the crop-weather relationship in the study area. Interestingly, the National Root Crop Research Institute, Vom and other agricultural research institutions in Nigeria and abroad have made notable achievements in increasing the yield of Irish potato. The crop breeders have developed varieties of Irish potato, which are capable of responding to improved cultural practices. They have gone a long way towards solving the pests and diseases problems both in the field and storage. In other to sustain these improvements, an appraisal of climatic parameters affecting Irish potato has become necessary.

In addition, there has been appreciable increase in the land area and output of Irish Potato in the area. This increase could be attributed to advances in agricultural technology such as the introduction and provision of extension services as well as way of diversifying sources of livelihood. However, despite the advanced techniques used in crop husbandry, the yield of Irish Potato is still variable over time and space in the area. Climate appears to be the major factor influencing the spatio-temporal variations in the yield of Irish potato in this area hence, the need for a research like this so as to unveil the prevailing situation.

STUDY AREA AND METHODS

This study investigates a crop-climate relationship using experimental approach. An experimental farm located at the Headquarters of the National Root Crop Research Institute Vom, in Jos-South Local Government Area of Plateau state served as source of data for the study. The site lies on latitude of 80 43'N and Longitude 80 46'E with an altitude of 1293.2m above sea level. Jos-South Local government area is one of the seventeen Local government areas in Plateau state of Nigeria. It is made of four districts; these include Vwang, Du, Gyel and Kuru. The Local government area has its Headquarters in Bukuru which is located south of Jos- North Local government. It is bounded by Barkin –Ladi Local government to the South, Riyom Local government to the South West, Jos-East Local government to the East and Bassa Local government to the West. The Local government has a population of 650,835 with an average land area of 1, 037km^2.

Two sets of secondary data were collected from the National Root Crop Research Institute Vom. The first set is agricultural data for Irish Potato yield in tonnes per hectare of the Nicola specie over a period of 20 years. The second set of data is the basic agro-climatic variables influencing Irish Potato yield over the same period for the months of April, May, June, July and August. These months constitute the growing season on the Plateau. The final yields of crop are in many cases dependent on satisfactory growth during earlier stages of crop development. In all nine climatic parameters collected during the growing season were analyzed with a view to estimating their contribution to yield of

Irish potato. These comprise three critical climatic data including rainfall; air temperature and soil temperature. Rainfall data were classified into total rainfall and rainfall amounts in April, May, June, July and August. Rainfall from April to August constitutes the rainy season in the study area. Temperature data were categorized into minimum and maximum temperatures. These data were analyzed and subjected to different statistical techniques, including correlation coefficient and multiple regressions. In order to select the agro-climatic indices that are critical to the yield of Irish Potato, a step- wise multiple regression analysis was adopted. The yields were expressed as dependent variables (Y) and climatic indices as independent variables (X). The general form of the multiple regression equation is presented as follow:

$Y = a + b1x1 + b2x2 + b3x3 + b4x4 + \ldots + bnxn$, where,

Y = Irish potato yield in ton/ha

a = constants

b = is the rise or falls as X increases

X1 = total rainfall (mm)

X2 = maximum temperature °C

X3 = minimum temperature °C

X4 = soil temperature °C

X5 = Rainfall in April

X6 = Rainfall in May

X7 = Rainfall in June

X8 = Rainfall in July

X9 = Rainfall in August

RESULTS AND DISCUSSIONS

Results and discussions in this study are considered under three sections. The first section treats the characteristics of the agro-climatic variables. This section was analyzed using descriptive statistics. The second section presents the relationship between agroclimatic variables and yield of Irish potato while the third section discusses the critical agro-climatic factors influencing the yield of Irish potato on the Plateau.

Characteristics of Agro-Climatic Parameters

The descriptive statistics of the agro-climatic variables were derived using SPSS statistical package. The mean total rainfall meets up with the requirement for Irish potato production. However, there is high variation between the annual data as seen in the results for range and variance. The average minimum and maximum temperatures of 24.82 °C and 18.85 °C respectively are within the range required by Irish potato crop. The standard deviation of 0.6 and variance of 0.4 in both cases means that there is less dispersion within the values. Soil temperature has similar pattern of variation like air temperature. With a standard deviation and variance of 0.6 and 0.4 respectively, it means that the element portrays a near uniform characteristics throughout the growing season on the Plateau.

Table. Descriptive statistics of Climate Variables and Yield of Irish Potato (1989-2008)

Variables	N	Range	Minimum	Maximum	Sum	Mean	Std. Deviation	Variance
Total Rainfall (mm)	20	196.80	575.30	772.10	13852.30	692.62	56.41	3182.42
Mean Max. Temperature (°C)	20	2.60	23.30	25.90	496.40	24.82	0.601	0.36
Mean Min. Temperature (°C)	20	2.00	17.80	19.80	376.90	18.85	0.59	0.35
Mean Soil Temperature (°C)	20	2.40	23.40	25.80	491.70	24.59	0.63	0.40
Mean Rainfall in April	20	55.60	40.30	95.90	1383.40	69.17	18.23	332.40
Mean Rainfall in May	20	70.30	126.10	196.40	3155.90	157.80	22.89	524.14
Mean Rainfall in June	20	50.80	129.30	180.10	3129.20	156.46	12.48	155.83
Mean Rainfall in July	20	99.20	213.10	312.30	5279.40	263.97	25.97	674.46
Mean Rainfall in Aug.	20	21.40	30.00	51.40	821.60	41.08	6.58	43.23
Yield (tonnes/hectare)	20	8.70	11.10	19.80	286.50	14.33	1.99	3.94
Valid N (listwise)	20							

There seems to be high variability only in April to August rainfall amounts between the years. This variation tends to be more in the months of July and May, while the month of August exhibits a lower variation in rainfall amounts within the 20 years considered.

It can therefore be concluded that the variations in the agro-climatic data presented above is partly responsible for the nature of variations in the yield of Irish potato in the area. The standard deviation and variance of 1.99 and 3.94 respectively, for the yield of Irish potato is relatively high. This calls for further probe into these factors so as to determine the degree of contributions of each of these factors. It is on the basis of this that the use of correlation and regression analysis was employed to determine the contribution of these factors to growth and yield of Irish potato.

Agro-Climatological Factors and Irish Potato

At the various stages of Irish potato development, striking findings were observed. For examples, the result of the correlation analysis between yield and the climatic parameters at sprouting to emergence stage revealed that maximum temperature and minimum temperature have significant correlations ($r = 0.594$ and $r = 0.548$ respectively) with Irish potato yield, at 1% probability level. This indicates that, temperature range of 21-26°C is good for sprouting of Irish potato. This consequently, plays a vital role in determining higher yield obtained on the plateau.

The high correlation obtained in this study confirms the assertion made by Ahmed that temperature ranges of 21-26°C is required for sprouting of tubers. He reported that a shift in temperature range of 21°C - 26°C at sprouting to emergence/vegetative stage may often induce knobbiness and secondary growth in tubers and consequently affects the yield negatively. At the tuber set initiation, total rainfall was found to correlate significantly with Irish potato at

5% probability level.This signifies the importance of moisture during the tuber set/initiation stage. It confirms the findings of Levi and Eliot that precipitation is significant and positively correlated at tuber set/initiation stage with Irish potato in Ireland.

The plants require more frequent supply of rainfall than many other tuber crops. It is the moisture and nutrient within this stage that is useful to the plant. Water is required for transpiration and regulation of leaves temperature. However, at the tuber bulking or ripening stage, minimum temperature was discovered to correlate negatively at 1% probability with Irish potato. This shows that the lower the temperature during this stage, the higher the yield will be.

This result confirms the assertions made by Levi and Eliot that minimum temperature of 15°C at tuber bulking stage positively correlates with Irish potato yield in Ireland. Lopez et al also stated that at bulking stage minimum temperature should not exceed 15oc for maximum yield, and that tuber decreases with increasing temperature at tuber bulking stage.

Furthermore, Kochalar reported that the minimum temperature of 15°C is required for tuber bulking, while Ochigbo observed that low temperature is more conducive to tuber growth at bulking stage and economic tuber production occurs when the average temperature falls below 15°C. More importantly, Ifenkwe and Okonkwo stated that, under irrigation, Irish potato production should coincide with the coldest month so that the time of tuber bulking will coincide with the period of low temperature.

Shimshe's finding corroborated this when he reported that once there is fluctuation in minimum temperature, it enhances rottenness and malformation in tuberization. Results for the total growing season indicate that Irish potato yield is significantly correlated with two of the variables at 1% probability level. Rainfall in April has high correlation with Irish potato yield.

This indicates that the higher the amount of rainfall in April, the higher will be the yield of Irish potato. It signifies the importance of rainfall during the sprouting to emergence/vegetative stage, which is a measure, to some extent, of the soil moisture available at emergence. During the raining season time of planting depends on the onset of rains. In Jos-South, Irish potato is planted when rain becomes stable usually between the last week of April and first week of May.

This result has collaborated Kowal and Andrew that the antecedent moisture status of the soil at planting is important for good germination or sustained growth of seedlings which may have effect on the final yield. Rainfall amount in July, with a correlation coefficient of -0.665, also its high relationship with Irish potato yield, though negatively, at 1% level of significance.

This indicates that, the higher the amount of rainfall in July, the lower the yield will be. This is not surprising because July coincides with tuber bulking/

ripening stage. The higher correlation obtained in this study confirms the assertion by Zaag et, al that excess soil moisture at tuber bulking/ripening stage causes poor soil aeration and root damage which influence crop development and yield. It further confirms the work of Nwakocha that blight causes between 40-80% reductions in yield.

The peak incidence is between July and August when the haulm of most susceptible varieties are destroyed by inciting pathogen phythopthora. The disease is accompanied by high relative humidity, dew and frequent rainfall.

Table. Results of Correlation of Climate Elements with Irish Potato Yield at Various Phonological Stages.

Climatic element	Sprouting to emergence/Vegetative	Tuber set/initiation	Tuber bulking/ripening	Total growing season
X_1=Total rainfall(mm)	0.332	0.470*	-0.122	0.333
X_2= max. Temp. °C	0.594**	0.202	0.412	0.250
X_3= min. temp. °C	0.548**	0.219	-0.616**	0.300
X_4= soil temp. °C	-0.041	-0.335	-0.097	0.362
X_5= Rainfall in April	-	-	-	0.599**
X_6=Rainfall in may	-	-	-	0.014
X_7= Rainfall in June	-	-	-	0.306
X_8= rainfall in July	-	-	-	-0.665**
X_9= rainfall in August	-	-	-	-0.391
** Significant at 1% r = 0.530		*Significant at 5%	r = 0.430	

Combine Effects of Agro-climatic Parameters on Irish potato Yield

The result of the bivariate correlation discussed above only depicted the isolated relationship between the climatic variables and Irish potato yield. It does not indicate the level of importance of such climatic variable influencing yield.

Therefore, in order to identify clearly those climatic variables that are critical to Irish potato production, those five climatic parameters that were found to have contributed significantly during the three stages of growth were further subjected to step-wise regression analysis.

Only those climatic parameters which contribute significantly to variation in Irish potato yield are included in the analysis. Two out of the nine observed climatic factors considered were found to have contributed significantly to the variation in the yield of Irish potato in Jos-South. These indices are rainfall in July and minimum temperature at tuber bulking/ripening stage.

The two variables together accounted for 58.1% of the total variance in Irish potato yield in the area. Rainfall in July has the highest contribution of 44.25% to yield variation and minimum temperature contributed 13.85% to yield variation.

Table. Step-wise Regression of the Agro-climate Parameters on Irish Potato Yield.

Variables	Partial R^2	Model R^2	T	P > t	F - ratio
X_8 = Rainfall in July	44.25	44.25	2.86	0.01	11.78
X_3 = minimum temperature at tuber bulking stage	13.85	58.1	2.37	0.03	

POTATO POTENTIAL YIELD BASED ON CLIMATIC ELEMENTS AND CULTIVAR CHARACTERISTICS

The potential yield is the maximum yield of a given species or cultivar possible achievable under the existing conditions of solar radiation flux density, with all the other environmental factors considered to be optimal. Therefore, the potential yield is determined by the biological properties of the cultivar and radiation resources available for utilization. This yield category practically expresses the solar radiation resources for cultivating a given genotype in yield units, whereas the commercial yield is the yield attainable under existing farm conditions that takes into account all the factors limiting the production process and the crop yield. Meteorological factors directly influence crop potential productivities, regulating its transpiration, photosynthesis, and respiration processes in such a way as to control the growth and development of the plants throughout their physiological cycle at a given site.

The interaction of the meteorological factors with the crop responses is rather complex. However, by means of studies on determination techniques for assessing physiological crop responses to environmental factors under field conditions it is possible to come up with mathematical models to estimate crop potential production as a function of meteorological variables with a good precision.

Several researches have been conducted aiming at quantifying the effects of the environment on growth, development and yield of many agronomic crops. Among the main environmental factors that strongly govern all physiological processes of the plants one should bear in mind global solar radiation flux density, air temperature, and available soil water content.

Potato yield improvements might be obtained by increasing the net daily photosynthetically radiation (PAR) through higher solar irradiance or longer photoperiod. The photoperiod duration doubles from December to June at 50oN, while PAR increases eightfold from 2.11 to 17.01 MJ m-2 day-1 due to higher elevation of the sun above the horizon with lengthening days. Gross carbohydrate production on standard clear days increases from 108 to 529 kg ha-1 day-1 at 50oN, whereas it remains at about 420 kg ha-1 day-1 yearround near the equator. Low solar irradiance is a yield constraint at 30 to 40oN in winter when potatoes are grown to escape the summer heat.

SARQUIS et al. stated that the magnitude of the effect of elevated temperatures on potato growth and final yield is determined by an intricate interaction between soil temperature, air temperature, solar radiation flux density, and photoperiod duration. Their data extended previous observations of reduction in photosynthesis rate under elevated temperatures.

Under field conditions reduced dioxide carbon assimilation rate could not explain the yield reduction observed; the temperature effect on assimilation was not as dramatic as it was on growth or yield. Other workers have reported a severe reduction in the rate of assimilation at air temperatures above 30°C under controlled experimental conditions.

In such cases, reductions in CO_2 assimilation rate were shown to correlate well with reductions in growth and yield. These contrasting results reveal the complexity of plant responses to the combined effects of water and temperature stress, which inevitably occur in association under field conditions. Knowledge of climatic requirements of potato and its physiological responses to the environment is extremely important to help growers produce high yields with good tuber quality under site-specific atmospheric conditions. Such knowledge aims to provide a scientific support to research programs related to crop weather modelling.

The SUBSTORPotato model takes into consideration daily air temperature, photoperiod, intercepted solar radiation, soil water and nitrogen supply. The model simulated fresh tuber yields ranging from 4 t ha-1 to 56 t ha-1 due to differences in weather patterns, soils, cultivars, and management practices. KADAJA and TOOMING proposed a relatively simple model POMOD to calculate potato yield, which permits generalization of the knowledge in different disciplines on the potato crop yield levels, using the measured physiological, ecological, agrometeorological, and agronomical parameters of the plant. The input variables of the model can be divided into four groups: daily meteorological information, annual information, parameters of location and cultivar.

The first group includes global radiation, air temperature, and precipitation. The location is characterized by geographical latitude and hydrological parameters. As to the cultivar factor, the parameters of gross and net photosynthesis, the coefficients of growth and maintenance respiration, and albedo of the crop are also needed. Simulation models for potato growth and yield were proposed by many researchers all over the world and are widely described in the literature. The LINTUL-POTATO simulation model establishes potential yield of a certain cultivar for a determined growing period and plant density, and is based on: incident photosynthetically active radiation (PAR), fraction of PAR intercepted by the crop, and radiation use efficiency to produce dry matter.

Phenological crop development is driven by accumulated air temperature, while development stage determines dry matter partitioning and, through haulm growth, the pattern of intercepted PAR is defined. The potential yield established with this simulation model was used by CALDIZ and STRUIT to perform a preliminary yield gap analysis regarding actual and attainable potato yield in different areas of Argentina.

Similarly to the potential productivity estimation model described by VILLA NOVA et al. and employed by VILLA NOVA et al. for sugar cane, we tested the performance of a model based on studies of maximum rates of carbon dioxide assimilation for a C3 crop species as a function of air temperature, a fraction of global solar radiation flux density (PAR), photoperiod duration and leaf area index to estimate the potential productivity of potato crop, cultivar Itararé, grown under adequate soil water supply conditions at four distinct sites of the State of São Paulo, Brazil. In order to assess the performance of the proposed mathematical model, the estimated values of tuber yield were compared to observed productivity data under irrigation conditions for the studied sites.

POTATO FARMERS FEEL EFFECTS OF CLIMATE CHANGE

Padam Damai grows potatoes in Kathmandu Valley, Nepal. The past year has been an extremely bad one for him and his family. Unusual snowfall in February and hailstorms in summer destroyed much of his crop twice in a row. Extreme weather events have become more frequent also in other potato growing regions of the world. In some areas of the Peruvian highlands, temperatures during the potato season in February dropped down to –30°C, causing major yield losses. In other regions, yields are diminished by drought. Are these events isolated phenomena or are they effects of climate change? Storms, droughts and floods have troubled mankind for as long as anyone can remember. However, in recent years many places have recorded a marked increase in extreme weather events – a development that scientists clearly attribute to climate change. Although farmers around the world have succeeded in continuously adapting to varying conditions, this frequent occurrence of extreme weather events is confronting them with serious problems. On the occasion of the International Year of the Potato, this issue of InfoResources Focus takes a look at the challenges that climate change poses to agriculture and, more specifically, to potato farming.

POTATO – IMPORTANT THROUGHOUT THE WORLD

As the fourth most important food crop after rice, wheat and maize, potatoes are of invaluable importance for the diets and livelihoods of millions of people worldwide. The potato embarked on its successful journey around the globe in the 16th century, when the Spanish brought it to Europe from the South American Andes. From here, the potato found its way to Asia in the 17th century, and to Africa in the 19th century. The crop's comparably short vegetation period allows farmers throughout a wide range of different climatic conditions to find an appropriate season for its cultivation. Global potato production has grown markedly in the past years, particularly due to increased production in developing countries. Improvements in crop varieties, seed potato and cultivation methods have led to higher yields. Moreover, a shift in eating habits in many countries towards more industrially processed potato-based products has boosted demand. In 2005, for the first time, more potatoes were grown in developing countries than in industrialised nations. The main producer is China, with a crop yield of 71 million tonnes, which amounts to over 20% of global production.

Ideal for Small Farmers

Potatoes belong to the crops that grow even in unfavourable conditions and at high altitudes. Moreover, a few other crops produce a comparably high yield of nutrients per cultivated area – a quality that is particularly welcome in regions where land is scarce. This is why potatoes are highly important for many farming families in the world's mountain regions. High yields and usually favourable sales prices enable manually producing small farmers to be competitive despite the fact that potato cultures require considerable labour input.

An Interesting Cash Crop

Potatoes are an important source of income for many farmers. In the Andes they are often the only cash crop grown by small farmers. In the tropical lowlands of Bangladesh and India they are cultivated mainly as an irrigated winter cash crop. Potatoes enjoy particular popularity among farmers in the highlands of Vietnam, who profit from favourable prices. They grow the tubers as a catch crop, in rotation with rice and maize, and while the income they earn from potatoes equals that from rice, it amounts to twice what they could generate from maize and sweet potatoes.

CLIMATE CHANGE AND ITS IMPACT ON

Along with the familiar difficulties related to pests and diseases, potato farmers are increasingly confronted with abiotic problems. Farmers and researchers report an increase in water stress, changes in rainfall distribution and intensity, hail, and increasingly frequent frost and snowfall at high altitudes. The growing frequency of extreme weather events is generally interpreted as clearly related to climate change. The newest report by the Intergovernmental Panel of Climate Change, published in 2007, states that global climate warming is an unequivocal fact. Projections by the IPCC predict a rise in global temperature by 1.8–4°C by the year 2100 due to the increase in greenhouse gases, depending on the scenario.

This is expected to have grave consequences for mankind and the environment. The critical threshold is said to be around a temperature increase of 2°C. Approximately 15% of the total worldwide greenhouse gas emissions are caused by agriculture.

An additional 11% result from deforestation, mainly for the purpose of gaining cropland. Carbon dioxide (CO_2) emissions in agriculture are chiefly caused by the use of fossil fuels during all kinds of agricultural activities, as well as tillage, burning of crop residues, and slash-and-burn deforestation. In addition, agriculture produces around half of the global methane emissions. These two greenhouse gases are many times more potent than carbon dioxide. The main sources of CH_4 are livestock production, irrigated rice cultivation, and storage of manure. N2O is released into the atmosphere through the soil following the inadequate application of artificial fertilisers and manure. By taking appropriate measures, agriculture has the possibility of reducing greenhouse gas emissions and thereby actively contributing to the mitigation of climate change.

Agriculture

Agricultural production is highly dependent on climate and weather factors. The impact of climate change on agriculture is accordingly high. In some regions, drought is leading to water becoming an even more valuable good, while other regions are affected by increasingly frequent flooding or even have to cope with both phenomena. All projections are based on simulations and therefore involve uncertainties. However, it is clear that climate change will affect different cultivation systems and crop cultures to a varying degree, depending on the region.

An important role is attributed to the time period within which changes are expected to occur. The consequences will not be felt equally soon in all regions. According to the authors of the working group II of the IPCC: *"Crop productivity is expected to increase slightly at mid to high latitudes for local mean temperature increases of up to 1–3°C, depending on the crop, and then decrease beyond that in some regions.*

At lower latitudes, especially in seasonally dry and tropical regions, crop productivity is projected to decrease even for small local temperature increases, leading to a growing risk of hunger.

Globally, the potential for food production is projected to increase with rises in local average temperature over a range of 1–3°C. Beyond, it is projected to decrease. An increased frequency of drought and flood occurrence is expected to have a negative impact on local crop production, especially in subsistence sectors at low latitudes."

Livelihoods and Food Security

These projections clearly indicate that the impact of climate change will be felt most in the countries of the South. In these regions, climate change will aggravate existing problems such as land degradation and water stress. The high dependency of large parts of these populations on agriculture increases their vulnerability to climate change.

Subsistence farmers, in particular, will be very directly affected by any yield loss. Many of these people do not have the financial means and the know-how required to adapt to changing conditions.

The exact extent of the threat climate change poses to human food security is difficult to predict and depends on the scenario: *"Under moderate scenarios, climate change appears to have a negligible effect on the numbers of people at risk to hunger."* In a first phase, yield losses in the developing countries will probably be compensated by yield increases in the countries of the North. However, the question arises of how food will be distributed among the world's regions. Especially the poorest people will hardly be able to afford food produced in the North.

Other sources indicate that a global temperature rise of 3°C or more will expose 600 million people to the risk of hunger. However, it is crucial to remember that climate change is not the only factor influencing future food security. Other important factors are population growth, trade, and the distribution of food, among other things.

Potato Farming

There are only few existing simulation models that can provide information on the influence of climate change specifically on potato farming.

Rising Temperatures: A Blessing and a Curse

A rise in temperature leads to increased transpiration in the plants, thus raising their demand for water. In many of the drier potato growing regions this will cause water stress, leading yields to decline. The effect will be further aggravated by changes in rainfall distribution. Where there is no possibility of irrigation in these regions yields will decrease even further, down to the extent where potato growing will become altogether impossible. For several countries, particularly in the tropics and subtropics, yield declines are expected to reach up to 20–30%. Overnight temperature has a crucial influence on starch deposition in potato tubers.

The ideal temperature range is between 15 and 18°C. Overnight temperatures above 22°C severely hamper tuber development. By contrast, the impact of climate change on potato yields is expected to be favourable in cultivation areas at higher altitudes. In many of these areas the climatic conditions for potato growing are improving due to the rising temperatures. This boosts yields and leads to an expansion of potato production to both higher altitudes and higher latitudes. In some regions it will become possible to grow potatoes as a winter crop.

However, the expansion of potato cultivation to higher altitudes also bears risks. High-altitude cropland is often situated on steep slopes, where growing potatoes can aggravate soil degradation due to the high degree of tillage involved. More research needs to be done regarding the effect that the stronger ultraviolet radiation at higher altitudes has on potatoes.

Higher Yields Due to Higher CO_2 Concentration?

The increased CO_2 concentration associated with climate change has a positive effect on crop yields when examined as an isolated factor. Under laboratory conditions this effect is particularly pronounced for potatoes and other tubers. Therefore, some researchers consider CO_2 fertilising a possible measure to curb yield losses due to climate change. However, the interaction between increased CO_2 concentration and higher temperatures is not yet sufficiently examined to make reliable statements regarding their effects on tuber development in potatoes. Newest findings suggest that under field conditions the positive effects of high CO_2 concentrations observed in the lab will prove to be considerably lower than previously expected.

Disease and Pest Pressure Might Increase

In some regions, disease and pest pressure on potato production will increase due to climate change. Late blight is expected to expand into areas that have previously been fairly safe from this disease. Moreover, it is very probable that in certain regions aphids will occur in greater number and in different seasons than so far due to more favourable climatic conditions. As virus vectors, aphids represent a threat particularly to seed production. At present, seed is cultivated at high altitudes and prior to the seasonal occurrence of aphids in order to keep it virus free.

Wild P otato Species in Jeopardy

The growing temperature increase puts additional pressure on the potato's wild relatives. 16–22% of all wild potato species are threatened with extinction by the year 2055. This is a dangerous situation, since wild relatives are important gene pools for breeding new varieties.

CLIMATE CHANGE AND LAND SUITABILITY FOR POTATO PRODUCTION IN ENGLAND AND WALES: IMPACTS AND ADAPTATION

In England and Wales, the potato industry has changed dramatically in recent decades, froma sector comprised of many small individual farms to one with far fewer but much larger agribusinesses, driven by the need to provide highquality product to the major processors and supermarkets. In 2009, more than 94000 ha of potatoes were cropped in England and Wales with an average productivity of 48 t/ha. Over half of that area was irrigated, mainly by hose reels fitted with rain guns or booms. The irrigation season typically extends from May to September.

Nationally, potatoes are the most important irrigated crop, accounting for 0·43 of the total irrigated area and 0·56 of the total volume of irrigation water abstracted. Although the volumes abstracted are relatively small, irrigation is concentrated in the drier eastern regions of England and peaks in the summer months, in the driest catchments when water resources are most scarce.

This creates conflict with other water demands, most notably those for public water supply and environmental protection. The shallow and sparse rooting system of potato plants, often resulting from soil compaction, makes it very sensitive to soil moisture stress. This provides little scope for error in terms of irrigation management, and losses in yield and quality can result even from brief periods of water shortage following tuber set. Thus, most rainfed production in England and Wales is concentrated on soils with higher available water capacities (AWC) in regions where summer rainfall is higher, such as Yorkshire, Lancashire and the West Midlands.

In contrast, potatoes grown on coarse-textured and well-drained soils are more susceptible to water shortage, and require supplementary irrigation to maintain yield and quality.

The viability of rainfed potato production depends not only on the pedo-climatic conditions but also on the cultivar being grown, its resistance to drought and the tuber quality required by the target market.

More than 170 potato cultivars are commercially grown in England and Wales and classified based on their planting and lifting date into 'earlies' and 'maincrop'.

Earlies are usually planted in mid-March or early April and lifted after 10–13 weeks, while maincrop potatoes are usually planted in the first half of April in southern England and in late April further north and lifted normally after 15–20 weeks. These dates vary from year to year depending on the weather conditions.

Since they have a shorter growing season, the yield of earlies is normally less than that of maincrop potatoes. Maris Piper is the most popular cultivar grown in England and Wales; in 2009 it represented 0·19 of the total potato cropped area.

The cultivars Estima, Lady Rosetta and Markies were also important. The cultivars Estima, Maris Peer, Harmony and Marfona are mainly grown for supermarkets prepacks, while cultivars Lady Rosetta, Hermes, Saturna and Pentland Dell are favoured for processing.

In the present study, the cropping characteristics used for modelling potato land suitability were based on data for a 'typical' Maris Piper crop, the most widespread maincrop cultivar grown by farmers to supply both processing and pre-pack markets; however, the modelling outputs are also relevant to other cultivars with broadly similar responses.

Stalham & Allen demonstrated that cultivars differed in their maximum depth of rooting as a consequence of the duration of active root growth rather than in the rate of extension over the first 3–4 weeks after emergence.

In a selection of 16 cultivars, Maris Piper rooted to 0·83 m, while the range was 0·67–0·92m. The other factor that led to the selection of Maris Piper as a test crop was that it is neither particularly drought-sensitive nor resistant and its yield response to irrigation is fairly typical of most cultivars tested. The cultivar is, however, fairly indeterminate, e.g. capable of maintaining full canopy for 50–80 days with moderate fertilizer nitrogen input (typically 150 kg/ha of total nitrogen).

Table. A summary on the planted area (ha), proportion irrigated, average yield (t/ha) and total production (t) of the top 10 cultivars grwon in England and Wales.

Cultivar	Cropped area (ha)	Proportion irrigated	Average yield (t/ha)	Total production (t)	Maturity
Maris Piper	19102	0·63	52·9	964652	Maincrop
Estima	9154	0·46	52·7	434758	Maincrop
Lady Rosetta	6491	0·60	52·3	332901	Maincrop
Markies	5867	0·54	51·6	297301	Maincrop
Maris Peer	4577	0·66	34·7	156660	Earlies
Marfona	4277	0·61	52·8	216216	Earlies
Saturna	3364	0·83	45·5	150518	Maincrop
Pentland Dell	2952	0·17	46·5	127195	Maincrop
Hermes	2914	0·28	50·0	144034	Maincrop
Harmony	2759	0·57	54·6	145919	Maincrop

Future changes in climate could affect potato production directly by impacting on plant growth and indirectly by influencing land management practices (e.g. ability of the soil to bear machinery for seed-bed preparation, and the planting and harvesting dates) and by providing a more favourable environment for new or existing pests and plant diseases. Warmer temperatures and elevated CO_2 levels are expected to result in more favourable growing conditions for most crops grown in northern Europe including potatoes, although of course there will also be some negative consequences, which will vary spatially and temporally. The impacts of climate change on the irrigation needs and yield of potatoes grown in England have been assessed by Daccache et al., who combined the downscaled outputs from an ensemble of general circulation models with a potato crop growth model to simulate future irrigation needs and yield under selected emissions scenario for the 2050s.

Assuming crop husbandry factors remained unchanged, farm yields were shown to increase marginally, while the average irrigation needs were predicted to increase by 14–30% under the 'most likely' low and high emissions scenario, respectively. However, these simulations were for specific locations and ignored future spatial changes in land suitability and the viability of rainfed production. Internationally, many studies have considered the impacts of climate change on future agricultural land use through scenario modelling and their consequent policy impacts, but there is remarkably limited literature on the impacts of potential changes in land suitability, a key factor influencing a country or region's ability to adapt agricultural practices to a changing climate. But such analyses can play a critical role in formulating future land policies given the multi-functional role of agriculture and its importance for ecosystem services.

For example, Hood et al. determined the future potential for growing cool season grapes, high-yield pasture and blue gum in Victoria, Australia by combining land suitability analysis with climate change scenarios within a geographical information system (GIS) framework. In Scotland, Brown et al. demonstrated the importance of soil moisture on land-use options, and how shifts in land-use potential have implications for both strategic resource planning and for adaptation actions. Their assessment highlighted not only potential changes in agriculture and other productive land uses but also repercussions for biodiversity and terrestrial carbon stocks.

As part of a broader study investigating the impacts of climate change on the UK potato industry, the objective of the present paper was thus to develop a methodology using pedo-climate functions and a GIS to model and map the current and future land suitability for potato production in England and Wales. The outputs would help the agrifood industry and its 3000 potato growers to identify regions where future rainfed potato production might become limiting, and where future irrigated production might be constrained due to water resource limitations.

MATERIALS AND METHODS

In summary, a three-stage methodology was developed. Firstly, the current land suitability for maincrop potato production was modelled and mapped using a GIS. This provided a reference or 'baseline' from which the derived land suitability classes could be validated against observed data on the spatial distribution of potato cropping (for rainfed and irrigated production). Secondly, future potato land suitability was modelled using the latest projections of climate change produced by the UK Climate Impacts Programme. A comparison with the baseline suggests how land suitability classes at each point might alter due to changes in rainfall, temperature and other agroclimatic variables, and thus potential impacts on current centres of production.

Finally, the relationships between land suitability and water resource availability were assessed, to identify catchments where future irrigated production might be most at risk of water constraints and whether it could relocate to catchments with less stressed resources. From this, the implications of climate change can be assessed, including where rainfed production might cease to be viable, which cultivars are likely to be most or least suited to changes in land suitability and the range of adaptation options (e.g. shifting production, new soil management techniques, drainage and new irrigation infrastructure) that should be considered. A brief description of each stage is given below.

Assessing Current Potato Land Suitability

Characterizing the edaphic and climatic regions suitable for the production of a specific crop type generally requires a long time frame, coupled with extensive experimentation and experience, and significant resources. However, provided that land types and local climates can be adequately specified and sufficient knowledge is available regarding crop responses to soil and weather factors, then land suitability models offer an alternative and rapid means of producing maps and data sets showing suitable areas for a particular crop. The Soil Survey of England and Wales (now incorporated in the National Soil Resources Institute) previously developed a suite of land suitability models for a range of arable crops. The present study used the land suitability model for potatoes, which defines a number of parameters of which most are climatic or soil related, but some are site-specific. A brief summary of the land suitability modelling framework used is given below.

Definitions of Land Suitability

The four land suitability classes used by the National Soil Resources Institute, termed well suited, moderately suited, marginally suited and unsuited follow the approach pioneered by the FAO. Well suited land has a high and sustainable production potential. There is adequate opportunity to establish the crop in average years at or near the optimum sowing time and harvesting is rarely restricted by poor ground conditions. Even in wet years, working conditions are acceptable and do not prevent crop establishment. There are sufficient soil water reserves to meet the crop water demand even in a dry year (typically defined as when reference evapotranspiration (ET) exceeds rainfall every month during the summer growing season).

Moderately suited land has moderate or high potential production, but yield is variable from year to year due to either shortage of soil water to sustain full growth or to poor conditions at crop establishment, affecting either sowing time or soil structure. Marginally suited land has a potential production that is variable from year to year with considerable risks, high costs or difficulties in maintaining continuity of output. The variability is due to climate interacting with soil properties. The criteria for unsuited land vary from crop to crop but are mainly related to climate, gradient and stoniness, although the latter can be largely overcome with de-stoning machinery. Clearly, near the climatic limits there will be favourable years which allow efficient production but other years could be too wet or too cool for the development of potato plants.

Classification of Suited Land

The two main criteria used for differentiating between the suited land classes are trafficability and droughtiness, as these can have a considerable impact on tuber quality as well as yield. The former is measured as machinery work days (MWD), as described by Thomasson & Jones, and the latter from available water (AW) and soil moisture deficit (SMD) data. A summary of each criterion is given below.

Trafficability

A number of thresholds for workability are definable in soil terms by moisture state, bulk density, penetration resistance, or shear strength. The most obvious threshold defines the boundary between 'go' and 'no-go' for vehicle movement on the land. Another threshold, perceived as less limiting, occurs when work on the land is not precluded but cannot be performed efficiently without damage to the soil. The need is to identify for the farmer 'no-go', 'acceptable' and 'optimum' conditions for tillage or other operations at any moment in time. Initially, Thomasson proposed assessing opportunities for cultivations in terms of 'good' machinery workdays using soil characteristics together with the rainfall pattern to predict the number of days when heavy machines have reasonable access to fields for crop husbandry practices.

Insufficient MWD can affect crop development indirectly. For example, a delayed planting date caused by heavy spring rains would shift the growing season and increase the cold damage risk in late development stages. The harvesting date may coincide with heavy rainfall with consequences for market price, yield and tuber quality.

In 2000, c. 0·20 of the potatoes grown in England were harvested late due to excessive autumn rain limiting machinery access onto land. Under the same climatic conditions, water logging and saturation are more likely to occur in finetextured and slowly permeable soils than in welldrained, coarse-textured soils.

The aim of the workability system adopted in the present paper is to estimate the average number of days acceptable for non-damaging traffic and tillage within the autumn period and in the spring period, using empirical relationships between soil and weather conditions. Threshold criteria for MWD can be generalized such that for autumn-sown crops an average value of more than 50 days is desirable to ensure that, even in a 1 in 4 wet year, there are at least 30 days to allow cultivation and drilling on major areas of land before conditions become unacceptable. For spring-sown crops, similar criteria would suggest a minimum of 20 days in average years to ensure that in the 1 in 4 worst (wet) year there is some opportunity to establish the crop.

Table. Land suitability classes based on MWD and droughtiness, for rainfed and irrigated potatoes.

MWD	Droughtiness			
(1 Jan–30 Apr)	>50/Irrigated	50 to 0	0 to −50	<−50
>30	Well	Moderate	Marginal	Marginal
20–30	Well	Moderate	Marginal	Unsuited
10–20	Moderate	Marginal	Marginal	Unsuited
<10	Marginal	Marginal	Marginal	Unsuited

MAPPING FUTURE POTATO LAND SUITABILITY

Uncertainties in the outputs from GCMs are divided between emission uncertainty and modelling uncertainty. As greenhouse gas emissions are determined by different driving forces such as demographic development, socio-economic development and technological changes, the Intergovernmental Panel on Climate Change has produced four different socio-economic storylines to describe the relationships between the driving forces and their evolution. Each leads to a different emission scenario, with c. 40 emissions scenarios and six marker scenarios being defined. The latest projections from the UKCIP, known as UKCP09, uses IPCC defined scenarios A1FI, A1B and B1, renamed for simplicity as high, medium and low, respectively. Modelling uncertainty results from incomplete understanding of the climate system or an inappropriate representation of the complex climate system within a single model. For this reason, many studies use the outputs of multiple future climate projections to increase the confidence in the climate projections. UKCP09 has dealt with the major source of modelling uncertainty by using the outputs of a large ensemble of variants from the HadCM3 GCM and from 12 other international GCMs as well as dynamically downscaling the GCM results using the regional climate model HadRM3. As a consequence, 10000 samples of possible and plausible changes for each climatic variable are available, and the results are presented in terms of likelihood probability.

In the present study, the central estimate for each climate variable for each pixel was extracted for each emission scenario as a monthly gridded data set at 25×25 km resolution, expressed as a percentage change relative to the baseline, and used to perturb the UK Meteorological Office 5×5 km resolution baseline climate data set. Consequently, the future climatic parameters such as rainfall, temperature and ET were obtained for England and Wales at a grid resolution of 5×5 km. Using the UKCP09 climatology, the individual criteria for land suitability were again integrated in the GIS and used to produce a series of maps showing the most likely future land suitability for each pixel, for rainfed and irrigated potato production, for the 2050s and 2080s, for the low- and high-emissions scenario, respectively. As before, the CORINE data set was used to constrain the spatial analysis to arable land only.

LAND SUITABILITY AND WATER RESOURCES

The reliance on supplemental irrigation is increasing in order to reduce the effects of climate variability on crop yield and quality. But water resources are under intense pressure due to rising demand, competition between sectors and the longer term threat of climate change. Concerns have also been raised over the potential impacts of irrigation on the environment. Indeed, in many catchments, summer water resources are already over-committed and additional summer licenses for surface and groundwater irrigation abstraction are unobtainable. Information on the distribution of potato holdings across England and Wales are collected annually by the Agriculture and Horticulture Development Board as part of its statutory duty.

The level of detail in the public domain depends on its commercial sensitivity, but the data can be used to map the spatial distribution of potato growers. In England and Wales, the water regulatory authority, the Environment Agency (EA), has assessed the availability of water resources for abstraction at a catchment level. Each catchment has been defined according to its resource status and allocated to one of four categories in order of increasing stress: (i) 'water available', where additional water is likely to be available at all flows including summer low flows, (ii) 'no water available', where no water is available for further licensing at low flows, although water may be available at higher flows with restrictions, (iii) 'overlicensed', where current actual abstraction is such that no water is available at low flows and if all existing licenses were used to their full allocation they could cause unacceptable environmental damage at low flows and (iv) 'over-abstracted', where existing abstraction is causing unacceptable damage to the environment at low flows, although water may still be available at high flows.

Using the GIS, these two data sets were combined to correlate the distribution of potato growers relative to water resources availability.

CLIMATE CHANGE EFFECTS ON THE GLOBAL RISK OF POTATO LATE BLIGHT

The risk of damaging levels of many crop diseases, referred to hereafter simply as 'disease risk', is strongly influenced by the weather. Infectious plant disease occurs due to the interaction of three factors: a favorable environment, a susceptible host, and a competent pathogen. Therefore, changes in weather due to climate change are likely to affect disease risk and the effects of plant disease on crop production. There is growing interest in plant disease risk under future scenarios and how to adapt disease forecasting models to new scales of application for scenario analysis. Potato late blight is an important crop disease caused by the oomycete *Phytophthora infestans* De Bary. Late blight is well-known for its role in the Irish potato famine and its current threat to potato production globally. Potato yield loss from diseases, animal pests and weeds was estimated to be around 40% of attainable production, with diseases alone accounting for 21% loss, and potato late blight is generally recognized as the most important potato disease. While late blight resistance is available, it is frequently not present in popular varieties so management is often dependent on pesticides, which can be a cost-prohibitive input for resource-poor farmers and can affect non-target species.

Models of the effect of weather on within-field risk of potato late blight have been evolving for almost a century, generally drawing on temperature and humidity as the most important predictors. Fry *et al.* created SimCast and Grünwald *et al.* further developed the SimCast model and demonstrated that it also performed well in a tropical highland location. The SimCast algorithm estimates the risk of damaging late blight levels, expressed as 'blight units', based on the temperature during the consecutive hours in a day when relative humidity is above 90%. SimCast thus uses hourly weather data as input. To rescale SimCast for larger time steps and easier application to wider geographical areas, we developed a metamodel of the relationship between weather (temperature and relative humidity) and late blight risk, based on SimCast output.

The metamodel mm*Monthly* uses monthly time-step temperature and relative humidity data to predict disease risk expressed as 'blight units'. Hijmans et al. evaluated contemporary severity of potato late blight indirectly using tactical decision models, Blitecast and SimCast to predict the need for a prophylactic pesticide application to control late blight in farmers' fields. These models were then scaled up to estimate the number of pesticide applications necessary to manage late blight globally. To achieve this, they used monthly climate data and a weather generator to temporally downscale the hourly weather data necessary for these models. The impacts of climate change on potato late blight have been studied in the Midwestern US and Finland, indicating the potential for increased risk.

Our goal was to provide the first global analysis of climate change effects on potato late blight. We used mm*Monthly* to evaluate the effects of climate change emission scenarios and the level of disease resistance on the change in disease risk. Furthermore, using the mm*Monthly* metamodel to evaluate risk illustrates the potential of metamodels for rescaling short-term forecasting models for scenario analysis. Results are discussed in the context of important agroecological areas and political boundaries. As an overview, we used gridded historic climate data and gridded, future predicted climate data. A crop model was used to estimate potato-growing seasons for each grid cell for reference and future periods. The climate data were then used in the mm*Monthly* model to evaluate late blight risk for a three-month potatogrowing season for each grid cell. The mm*Monthly* model estimates blight units, a measure of the relative risk of damaging levels of potato late blight occurring given an input of monthly temperature and monthly relative humidity. Because late blight occurs so widely in potato production systems and the driving factors are the weather conditions, temperature and relative humidity, we assumed that inoculum was not a limiting factor in areas where potato is grown.

We used CRU CL 2.0 grid mean monthly temperature, mean monthly relative humidity and mean monthly precipitation data from New *et al*. as our reference climate data. These data are at a spatial resolution of 10 arc minutes covering the time period from 1961 to 1990. Future climate emission scenario data were downloaded from the World Climate Research Programme's Coupled Model Intercomparison Project phase 3 multi-model data, that we statistically downscaled to 10 arc minute resolution. Only data for General Circulation Models (GCM) that provided maximum temperature, minimum temperature, and vapor pressure were selected, relative humidity was calculated from these data. Ensemble model averages of the non-weighted means of the GCM outputs for each respective climate scenario were created for mean monthly temperature and mean monthly relative humidity. The global daily average blight unit accumulation per month was calculated from mean monthly temperature and mean monthly relative humidity using the mm*Monthly* metamodel for the 1961-1990 reference climate normal, and three future 20-year time slices, (referenced hereafter by the midpoint year of the time slice: 2000-2019, 2040-2059 and 2080-2099. The relative risk of damaging levels of late blight was then averaged using a three-month moving window to provide the average daily blight unit accumulation for 12 three-month time periods representing threemonth potato growing seasons. Optimal potato planting dates were estimated using the ECOCROP model as implemented in the dismo package in R using mean monthly temperature, mean maximum monthly temperature, mean minimum monthly temperature and mean monthly precipitation data.

The first day of the first month of a three month growing season in which planting would produce the highest potato yield for each grid-cell was calculated. Optimal planting dates were generated for the reference climate normal and the ensemble GCM outputs for each of the three respective future time slices. These data were used to estimate late blight risk for what would be a geographic location's most productive potato growing season in the absence of pests and disease. Total rainfed and irrigated potato production by country was used to remove areas where potato production is currently limited. Several countries were selected for further analysis because they are representative of highland or lowland tropical potato production, areas where potato is an important crop for poverty alleviation and where late blight is difficult to manage because of year-round potato production. These countries include Colombia and Ecuador in the Andean highlands; Ethiopia; Rwanda in the Lake Kivu highlands region; Nepal in the Himalayan highlands and Indo-Gangetic Plain, and Indonesia in the South East Asian highlands.

WHAT CAN POTATO FARMERS EXPECT FROM THE FUTURE?

Will potato farmer Padam Damai be able to keep up his production, and will his children be able to grow potatoes in 30 years? Despite the numerous model calculations and projections we currently still know much too little about climate change and its impact on agriculture. Time will show how potatoes will actually react to the changing climatic conditions. Farmers who are cultivating potatoes in regions where conditions are critical even today, may no longer be able to do so a few years from now. They will have to find other crops or engage in off-farm activities. Most of them will be forced to adapt.

They have already been able to implement some adjustments or will be able to do so in future without outside help. For other measures they need external support. At the policy level, mainstreaming climate change into national programmes and strategies is highly important as a prerequisite for timely assessment of vulnerabilities, as well as planning and implementation of measures. There are no simple recipes for the implementation of these strategies. Each country requires the strategy best suited for this country's specific situation. Many of the least developed countries (LDC) have formulated such strategies in their National Adaptation Programmes of Action.

These programmes identify vulnerabilities to climate change and point out first urgent adaptation measures. In agriculture, and particularly in potato production, much depends on successful breeding of new varieties. If researchers succeed in breeding more stress resistant varieties that produce sufficient yields even under changed conditions, potatoes will continue to be able to be grown in many regions also in future.

One issue will become increasingly important as water stress increases: water productivity. This characteristic refers to the amount of "crop per drop" a plant produces. With yields of 6.2–11.6 kg per m3, potatoes have a higher water productivity than rice, wheat, and maize. The International Potato Center (CIP) sees a chance that potatoes will continue to make an important contribution to food and income security in future: *Given the potato's high productivity per unit of land and time and value, as both a staple and a cash crop, increasing stress tolerance in potato has a great potential to contribute to food and income security, mitigate poverty, and reduce farmers' risk in vulnerable agricultural environments.*

5

Soil Management for Potatoes

INTRODUCTION

SOIL: A FARMER'S "SILENT PARTNER"

Soil is the basis of farming. It delivers water and nutrients to crops, physically supports plants, helps control pests, determines where rainfall goes after it hits the earth, and protects the quality of drinking water, air, and wildlife habitat. The goal of soil management is to protect soil and enhance its performance, so you can farm profitably and preserve environmental quality for decades to come.

Why Learn More About Soil Management?

Consider the valuable services your soil provides:

- Growing crops. Soil delivers nutrients and water and gives plants structural support. Could your soil cycle nutrients more efficiently so you save on input costs and your crops are healthier? Could your soil store more water so crops do better during dry spells?
- Controlling water flow. Soil helps control how water moves over and through the earth's surface. Does rainfall quickly fill waterways rather than moving slowly through your soil? Could you reduce the amount of organic matter, nutrients, and soil you are losing to erosion?
- Filtering water. Healthy soil can filter and decompose organic substances such as manure, agricultural chemicals, and other compounds that can pollute air and water.
- Storing carbon. Soil is a storehouse of carbon. As concern grows about increased atmospheric carbon dioxide, farmers may receive financial benefits for increasing the amount of carbon stored in their soil.

You Determine How Soil Performs

You cannot control slope, texture, climate, and other critical soil factors. But you can control tillage, crop rotations, soil amendments, and other management choices. Through these choices you change the structure, biological activity, and chemical content of soil, and you influence erosion rates, pest populations, nutrient availability, and crop production.

SOIL MANAGER

SIX PRACTICES THAT IMPROVE SOIL PERFORMANCE

Improving soil performance requires different actions on each farm. Most soil-friendly farm practices fall into one of six groups. Each of these practices is further explained in other publications in the series.

1. *Adding organic matter:* Regular additions of organic material may be the most important way to enhance soil quality. Organic matter improves soil structure, enhances water and nutrient holding capacity, protects soil from erosion and compaction, and supports a healthy community of soil organisms. Organic matter includes residue and roots from the previous crop, animal manure, cover crops, or amendments from off the farm.
2. *Avoiding excessive tillage and soil compaction:* Tillage is valuable for loosening surface soil, preparing the seedbed, and controlling weeds and pests. But tillage can also break up soil structure, speed the decomposition and loss of organic matter, increase the threat of erosion, destroy the habitat of helpful organisms, and cause compaction. Reducing tillage minimizes the loss of organic matter and increases the residue protecting the soil surface. Compaction reduces the amount of air, water, and space available to roots and soil organisms. Compaction is caused by traveling on wet soil or by heavy equipment.
3. *Managing pests and nutrients efficiently:* In this century, pesticides and chemical fertilizers have revolutionized U.S. agriculture. In addition to their desired effects, they can harm non-target organisms and pollute water and air if they are mismanaged. Nutrients from organic sources also can become pollutants when misapplied or over-applied. Efficient pest and nutrient management means applying only the necessary chemicals, at the right time and place to get the job done; testing and monitoring soil and pests; and adding non-chemical approaches to your management toolbox (such as crop rotations, cover crops, and manure management).
4. *Keeping the ground covered:* Bare soil is susceptible to wind and water erosion, and to drying and crusting. Groundcover protects soil, provides habitats for larger soil organisms (such as insects and earthworms), and can improve water availability. Farmers often leave crop residue on the surface to cover the ground between growing

seasons. Living cover crops create new organic matter and help feed soil organisms. Groundcover must be managed to prevent problems with delayed soil warming in spring, diseases, and excessive build-up of phosphorus at the surface.

5. *Increasing diversity:* Diversity is beneficial for several reasons. Each crop contributes a unique root structure and type of residue to the soil. A diversity of soil organisms helps control pest populations, and a diversity of cultural practices reduces weed and disease pressures. Diversity across the landscape can be increased by using buffer strips, small fields, or contour strip cropping. Diversity over time can be increased by adding crops to the crop rotation or by varying tillage practices. Changing vegetation across the landscape or over time not only increases plant diversity, but also the types of insects, microorganisms, and wildlife that live on your farm.
6. *Monitoring soil performance:* Nothing can replace the value of "casual" observations of how your land is changing from day to day and year to year. Yet, to fine-tune management practices and promptly determine whether changes in soil or crops are significant, you also need to make systematic observations of the soil.

THE SOIL SCIENTIST

This section gives background information about how soil works. Understanding the basics will help you apply management recommendations found in this and other publications in this series.

WHAT IS SOIL MADE OF?

It takes thousands of years for rock to develop into soil, and hundreds of years for rich organic layers to build up. Soil is made of air, water, mineral particles, organic matter, and organisms. Half of soil is pore space. Generally, pores are about half filled with water and half air, though the proportion varies greatly depending on weather, plant water use, and soil texture. Most of the solid portion of soil is mineral particles. Organic matter may make up only 5% to 10% of the volume of soil, but it is critical in holding soil particles together, storing nutrients, and feeding soil organisms.

Mineral particles are divided into three groups based on their size: clay, silt, and sand. Soil texture depends on the proportion of particles from each of these groups. For example, a loam has similar proportions of all three classes of particles. A sandy loam is higher in sand; a clay loam is higher in clay.

Soil structure, or how soil is put together, can be as important as what it is made of. Most soil particles are held together in aggregates of many particles. The size and stability of these aggregates determine the size of pores. Soil texture is difficult to change, but farming does impact soil structure.

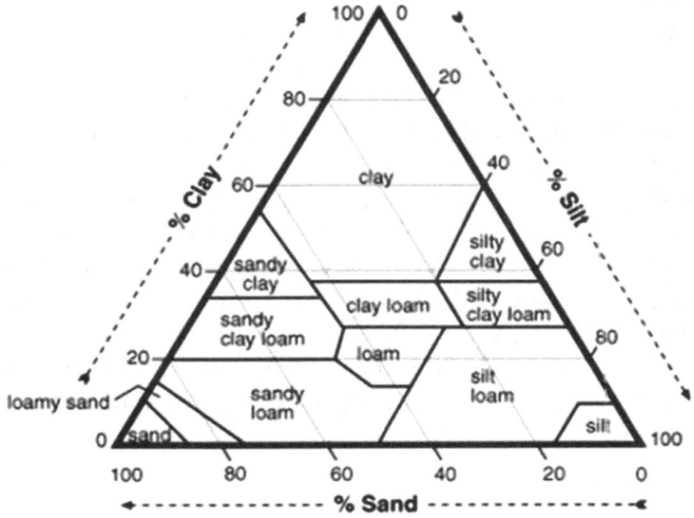

Soil texture triangle particle diameters
clay: < 0.002mm
silt: 0.002mm to .05mm
Sand: .05mm to 2mm

HOW DOES SOIL WORK?

Regardless of the kind of farming you do, the rules of soil chemistry, physics, and ecology are the same. Four processes make your soil what it is: formation of soil structure, nutrient cycles, water cycles, and life cycles of soil organisms.

Soil structure

Soil particles range in size from gritty sand particles as large as 2 mm to microscopic clay particles 1000 times smaller. These particles rarely exist separately in the soil. They normally combine into clumps called aggregates or peds. A few particles bind into tiny microaggregates. Microaggregates, in turn, combine to form larger aggregates. Ideally, soil will have a wide range of aggregate sizes and pore sizes. Aggregates in healthy soil will be stable and resist breakage when tilled, hit by rain, or otherwise disturbed.

How soil structure develops

*How do aggregates form?*Several biological, physical, and chemical processes interact to form aggregates and then stabilize them. Microbes decomposing organic matter create compounds that cement soil particles together. Synthetic polymers and sugars excreted from roots can have a similar cementing effect. Humus - the stable organic compounds - also binds soil particles together. Fungal hyphae and fine roots surround and stabilize aggregates. This is why surface residue and other organic matter improve soil structure - the residue is food for fungi and bacteria that in turn help form and stabilize soil aggregates. Larger organisms (such as insects and earthworms), enhance soil structure when they burrow through the soil and deposit fecal pellets that become stable soil aggregates. Physical and chemical processes are also important to the formation of soil aggregates - especially smaller aggregates. Particles are physically pushed closer together by freezing and thawing, wetting and drying, and roots pushing through the soil. At the molecular level, electrochemical charges bond clay particles together. In arid areas, this process may be disrupted by large amounts of sodium (Na^+) ions. In tropical and semitropical soils, iron oxides can cement together aggregates.

*How do surface crusts form?*Rain drops break apart soil aggregates and then fine clay particles clog the spaces between aggregates and form a crust on the soil.

Improving soil structure

Several of our six soil-friendly practices impact soil structure:

- Organic matter management. Regular additions feed the organisms that build soil structure.
- Tillage practices. Over time, tillage increases the decomposition of soil organic matter and breaks up aggregates - especially when tillage is done in wet soil. Residue left at the surface protects surface aggregates from rain and encourages the growth of fungi that help stabilize aggregates.
- Compaction prevention. Compaction pushes aggregates together and eventually breaks them down.
- Crop choices. The dense roots of grasses, small grains, and pastures stabilize soil aggregates and improve structure.

Nutrient cycles

Soil is the storehouse for the nutrients that plants need. It is a dynamic environment in which soil organisms, chemistry, and physics are continually acting to change the form of plant nutrients. Plant roots draw most of the nutrients they need from the soil solution - the water and dissolved minerals in soil pores. The amount of nutrients in the soil solution at any one time is just a fraction of that needed by plants over the course of the year. The soil solution must be continually replenished from the store of nutrients in minerals and organic matter. Many farmers also contribute nutrients to the soil solution by adding readily available fertilizers once or more each year. Soil particles and plant residue are made of large quantities of nutrients that are unavailable to plants. Eventually, the soil particles weather into sand, silt, clay molecules, or mineral ions. Plant residue eventually decomposes into mineral ions or reforms into humus.

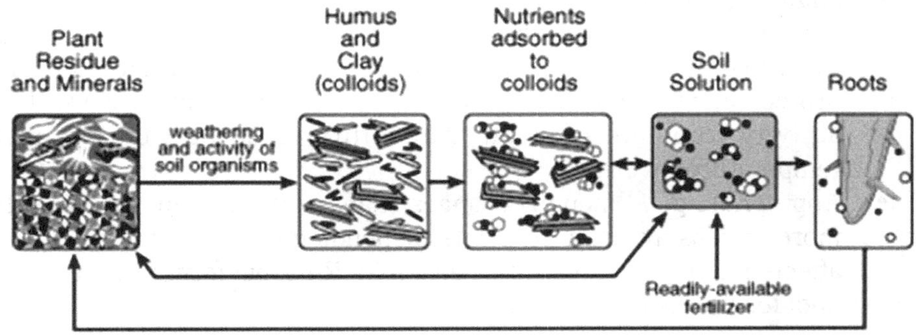

Plant nutrients exist in several forms

Clay and humus are not absorbed by plants, but they hold nutrients on their surfaces. The amount of places available on clay and humus to hold nutrients is called the "exchange capacity" of the soil. Cation exchange capacity is the number of places available - the storage capacity - for positively charged ions, including calcium, magnesium, sodium, and potassium ions. Nutrient ions attached to the exchange sites on clay and humus are released into the soil solution for use by plants and soil organisms such as bacteria and fungi. As plants draw nutrients out of the soil solution, more may be released into solution from exchange sites on clay and humus. Fertilizers added to the soil solution do not remain unchanged, waiting to be used by plants. Like minerals from other sources, they become attached to exchange sites, are used by microorganisms, and are transformed.

Roots get nutrients in three ways:

- The root grows into an area where the soil solution has not been depleted of nutrients;
- After a root depletes the nutrients near it, nutrients will diffuse into the deficient area; and
- Water flows towards the root and carries nutrients. This means that prolific root growth and adequate water are essential to plant nutrition.

What is soil fertility?

Where is the most fertile land on your farm? What makes it more productive than other fields? The soil may test high in nutrients, but it probably has additional characteristics that make it your best land. Soil fertility is not just the amount of nutrients, but whether plants can get the nutrients when they need them. In other words, a fertile soil will have:

- Good rooting environment. To grow and find nutrients, roots need well-drained soil with a crumbly, uncompacted structure.
- Adequate water. Soil with good "tilth" will have good water infiltration and water-holding capacity.

- High organic matter. Organic matter is a source of many nutrients, improves the rooting environment, and helps hold water in the soil.
- Active soil community. Soil organisms release and retain nutrients, protect plants from pests, and even enhance plant growth. Their activity depends on food availability, pH, and moisture and temperature levels.
- Appropriate pH. When pH changes, many nutrients can become either more or less available to plants, depending on the nutrient. pH also affects microbial activity. For example, Rhizobia form nitrogen-fixing nodules poorly in acid soils.

Improving nutrient cycling

Several of our six soil-friendly practices are important to nutrient cycling:

- Organic matter management. Organic matter supplies nutrients for plants and feeds the soil organisms responsible for cycling nutrients. It is a "slow release" fertilizer.
- Tillage practices. Tillage triggers the decomposition of organic matter and the release of nutrients, and mixes nutrients throughout topsoil. Excessive tillage reduces organic matter and the nutrient-holding capacity of your soil.
- Compaction prevention. Preventing compaction improves the ability of roots to grow through soil to reach available nutrients.
- Fertilizer management. You have choices about which form and how much of a nutrient to use, and when and how to apply it.
- Crop choices. Each crop affects nutrient cycling differently and encourages a different mix of soil organisms. Deep-rooted crops draw up nutrients that other crops cannot reach. Legumes can add nitrogen to the soil.

Water cycle

If you have watched crops suffer through a drought or have seen yields jump after installing drainage tiles, then you understand that water is a fundamental factor for good crop yields. Like animals, plants need large amounts of water every day. Yet, too much water deprives roots of air and makes soil susceptible to compaction. Water carries soil, nutrients, and other substances, so water management is essential to controlling erosion and pollution.

Three major water processes occur in soil: water gets into the soil, is held by the soil, and drains out of the soil. How these processes occur depends on soil type and management. Infiltration is the rate at which water gets through the surface and into the soil. With higher infiltration, more water will be available to plants and less will run off the surface, erode soil, and wash away nutrients. Crop residue, living plants, or a rough soil surface will slow down the flow of water so more can infiltrate. A soil crust reduces infiltration and can be minimized by leaving surface residue, improving organic matter levels, and enhancing biological activity. Available water holding capacity is the amount of water soil can hold for plant use.

After water gets into the soil, the surface tension of water holds it in soil pores against the pull of gravity. Because of surface tension, small pores in fine-textured soils such as silt and clay loams hold more water than the large pores in sandy soils. Organic matter also holds large quantities of water. The maximum amount of water that a soil can hold against the pull of gravity is called the field capacity. Generally when a soil is at field capacity, half the pores are filled with water.

Plants cannot use all of the water in soil. As water evaporates and is drawn out of soil by plants, the water content gradually declines until plants can no longer extract the small amount of remaining water held tightly to soil particles. This remaining amount of soil water is called the wilting point. Clays have a high wilting point. They hold water more tightly than coarser-textured soils and so less of the water is available to plants.

Thus, although silt loams hold a bit less water than clay loams, more of it is available to plants. Salts prevent roots from absorbing water and thus increase the wilting point. Drainage or percolation is excess water that soil cannot hold that moves out of the rooting zone so roots and organisms can get air. After a heavy rain, the soil will be saturated (all the soil pores are filled with water). Many roots and soil organisms will die if the excess water does not quickly percolate out and allow air into the pores again.

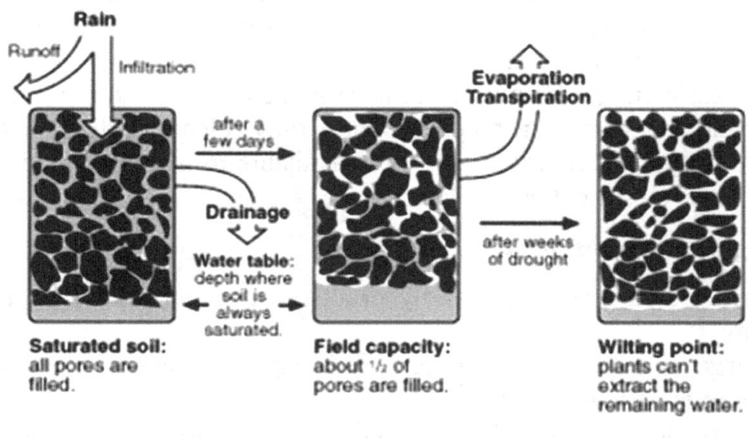

Changes in Soil Water. Field capacity minus wilting point is the amount of water available to plants.

Improving the availability of water

Several of our six soil-friendly practices increase field capacity and improve infiltration:

- Organic matter management. Organic matter significantly increases the water-holding capacity of soil in two ways. It absorbs and holds large amounts of water, and it improves the structure of the soil - increasing the total volume and size of pores that can hold water and preventing soil crusting.
- Tillage practices. Leaving residue on the surface slows runoff and prevents crusting. Residue encourages populations of earthworms and other burrowing organisms and water infiltrates rapidly into their burrows.
- Compaction prevention. Compaction reduces water holding capacity by reducing the number and size of soil pores.
- Erosion control. Erosion reduces the depth of your soil and its water-holding capacity.

Life Cycles of Soil Organisms

Soil organisms were mentioned in connection with all three of the previous soil processes. Their life cycles are an integral part of the formation of soil structure and nutrient cycles, and their activity is highly dependent on the water status of soil. Temperature, food supply, and pH are other factors that determine what lives in your soil and when they are active.

The food web of organisms living underground is at least as diverse and complex as the ecosystem of plants and animals living above ground. Practically all the energy for the food web comes from the sun.

Plants and other photosynthesizers convert the sunâ□™s energy and carbon dioxide into the carbon compounds used by other organisms. Plant-eaters create compounds that other organisms need.

As bacteria, fungi, earthworms, microscopic insects, and other organisms consume and transform carbon compounds, they release carbon dioxide back into the atmosphere and make nutrients available to plants. Farmers depend on these life cycles for their livelihood.

Improving the soil ecosystem

Several of our six soil-friendly practices can improve your soilâ□™s ecosystem:

- Organic matter management. Soil organisms need a regular supply of organic matter for food.
- Increase diversity. It is generally beneficial to increase the diversity of organisms in the soil. Diversity is increased by using different management practices, including long crop rotations, changing tillage practices from year to year, supplying several types of organic matter, or creating buffer strips or other vegetation that adds variety to the landscape.

SOIL LIFE

A YEAR IN THE LIFE OF YOUR SOIL

Dates are approximate for Minnesota and vary each other.

	Spring	Summer	Fall
Structure & Temp.	Saturated soils have a weak structure and are prone to compaction. Spring soil is highly susceptible to wind and water erosion, because soil is often bare and soil structure is weak. By late spring, bacterial and fungal activity will help stabilize soil aggregates, and plant growth will begin to protect the soil surface.	Even healthy soil is difficult to penetrate when it gets very dry.	Risk of soil compaction increases again as soil moisture levels rise and heavy harvest equipment enters the fields.
Nutrients	Bacteria and fungi are becoming active, but not necessarily enough to provide nutrients to young plants. Side-dressing of nutrients may be useful.	If a young forage is plowed down, soil organisms quickly release N and other nutrients to the next crop. A low-N amendment, such as straw, may trigger N deficiency in a crop, as soil organisms use soil nitrate to decompose the high carbon amendment.	Fall is a good time to test soil - leaving all winter to use test results for planning the next cropping season. Cool season cover crops can take advantage of nutrients not being used by the main-season crop, and can prevent nitrogen losses.
Water	Compacted soils drain slowly. The loose, rough surface of newly tilled soil readily allows rain water to infiltrate into the soil - unless a crust forms.	High temperatures in July and August cause high water loss through evaporation and transpiration from plants. Soil water stores are at their lowest.	Soil water levels begin to recharge as temperatures drop and evaporation and transpiration slow.
Organisms	Warming temperatures, air from tillage, and food from a plowed-down crop trigger a high level of activity. Organisms give off heat, carbon dioxide, and nutrients as they consume residue.	Any change in environment - new food, hot spell, heavy rain - will change microbial activity. In late summer, organisms from bacteria to earthworms are less active and reproduce less because of the lack of water.	Microbial activity increases in response to greater soil moisture and new food in the form of roots and residue from harvested crops.
Plants	Tillage triggers weeds to sprout. As crop roots begin to grow, they continually contribute organic matter to the soil. They exude organic compounds and slough off dying cells.	Root growth slows as plants produce flowers and other reproductive growth. By this time, roots may have extended deep into subsoil in search of water and nutrients. Warm season plants such as corn and some grasses flourish in the heat .	Over the course of the year, 40% of the photosynthetic energy captured by a plant is used by the roots, and enters the soil system. Most of this carbon is released from the roots as CO_2.

SOIL POTASSIUM FOR POTATOES

Fertilizer experiments—and a nutrient survey during the 1958 spring season- showed that some fields in Kern and Tulare counties are being depleted of exchangeable soil potassium to the point of deficiency, through frequent cropping to potatoes. Potatoes as a crop are one of the heaviest known feeders on potassium. Nutrient removal studies in the Kern area showed that a 4M-sack per acre crop of potatoes removed from the soil more than 250 pounds of potash—K,O-per acre, while over 200 pounds were removed by the tubers alone. With such large quantities of potassium being removed annually by the potato crop, available potassium in the soil is reduced considerably, unless weathering of the potassium minerals occurs at rates sufficient to replenish the supply. To determine the extent of soil potassium depletion by frequent potato cropping, eight experiments were conducted in certain areas of Kern and Tulare counties. The plots selected were where previously observed foliar symptoms and plant tissue analyses indicated that a response to potash fertilization might be realized. An additional 20 fields in Kern County were chosen at random for making a nutrient survey of White Rose potatoes.

Table. Soil and plant potassium and yield from Eight field experiments in Kern Area-1958

Field	Soil pH	Exchange-able Potassium (K_2O) ppm	Percent Potassium (K) in Petiole At the last sampling		Yield cwt/acre	
			No potash	200 lbs. K_2O	No K_2O	200 lbs. K_2O
Cropped	6.0	101	7.8	10.6	211	230
Uncrop.	7.8	193				
Cropped	4.6	106	4.8	8.7	156	175
Uncrop.	6.7	275				
Cropped	4.6	104	7.4	9.2	195	217
Uncrop.	6.6	180				
Cropped	4.7	105	4.5	6.4	194	206
Uncrop.	6.9	287				
Cropped	6.9	70	0.6*	5.8	174	224
Uncrop.	6.7	201				
Cropped	7.0	46	1.1*	5.4	194	266
Uncrop.	8.0	132				
Cropped	7.1	101	4.2	7.6	296	304
Uncrop.	7.8	435				
Cropped	7.4	93	2.4*	5.2	370	391
Uncrop.	6.0	149				

*Considered to be in range of K deficiency (less than 4% K at late season).

Relationship between percent potassium in patat0 leaf petioles and exchangeable potassium in the soil.

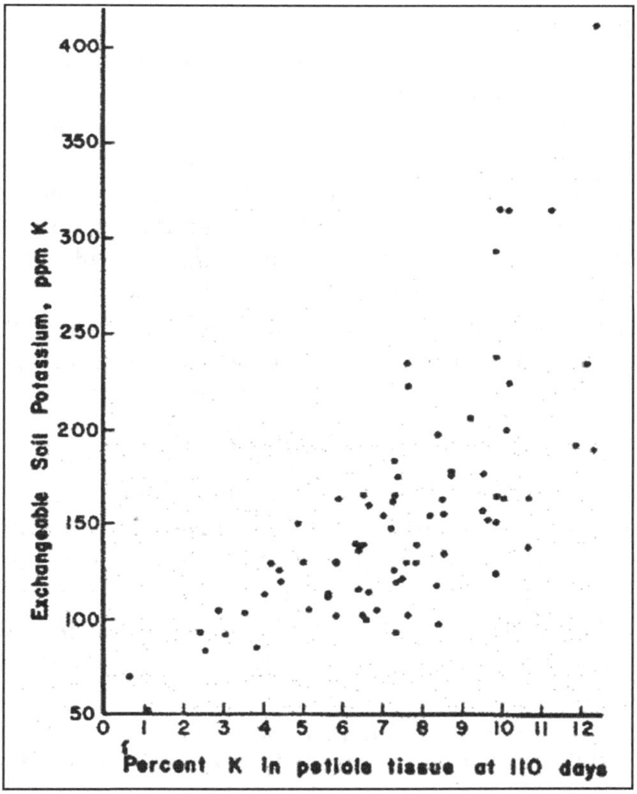

Samples of soils from the experimental plots and survey fields were taken from the plow layer prior to planting and fertilizing. Soil samples were collected also from adjacent areas which had not been cropped. Comparisons of the exchangeable potassium content of the cropped and uncropped soils as shown in the accompanying tables indicate that as a result of cropping the potassium available to plants has decreased in some soils to levels as low as one fourth that in the adjacent uncropped soils. Although these po'assium differences between cropped and uncropped soils were appreciable in most *of* the 20 survey fields, they were generally of lesser magnitude than the differences in the eight experimental fields which were known to be deficient in potassium. The 1958 studies-and previous experiments- demonstrated that when exchangeable soil potassium is be1o.v 100 ppm-parts per million-it is in the deficiency range for potatoes.

In these studies exchanqeable soil potassium was determined by leaching 50 grams of soil with 500 milliliters of neutral, normal ammonium acetate and analyzing the leachate for potassium. In three of the eight field experiments, potassium was deficient and in the other five fields it was bordering on deficiency. Plant tissue analyses of petiole samples from the same fields verified the fact that potassium was deficient in the three fields and low in the remaining five. Yield responses to potassium fertilization were observed in all eight fields with the greatest responses ik fields where exchangeable soil potassium was lowest. Some of the older fields of the 20 in the survey were on the borderline of potassium deficiency for potatoes as indicated by plant and soil analyses. In two h were more recently brought ovation and where steer manure had been applied before planting, the exchangeable soil potassium was in excess of *300* ppm in the cropped soil and over 400 ppm in the uncropped. These were the only soils of the entire stud) where potash application failed to bring about a marked increase in the foliar potassium content of the potato plants. Even in soils which were considered to have adequate levels of exchangeable potassium-above 150 ppm-potash fertilization resulted in higher levels of potassium in the petioles. In all fields considered there was good granis of carbon dioxide per kilogram per hour, and at 68°F was 190 rnilligranis of carbon dioxide per kilogram per hour, a fivefold increase over the temperature range studied. At higher temperatures, rates dropped off rapidlyduring the first six days in storage* and increased again toward the end of the storage period. a phenollienon associated with the appearance of decay organisms.

At lower temperatures the rate of respiration declined gradually for the first 10 days. and leveled off for the remainder of the holding period. The number of days before half of the sprouts in a sample reached a quality rating of 3-411-salable-is indicated by the termination of each curve in the graph.

Table. Soil and plant potassium from 20 Potato Nutrient Survey field in Kern County-1958

PPM exchangeable Potassium (K)		Percent Potassium in Petiole[1]	
Cropped soil	Un-cropped soil	No K applied	200 lbs. K_2O applied[2]
100	160	5.8	8.4
103	178	2.8	6.4
108	150	7.8	8.3
112	150	3.7	5.2
119	198	7.1	11.7
128	130	6.4	7.2
131	121	7.6	8.8
133	315	7.9	9.5
145	240	6.6	9.7
147	189	5.9	9.0
160	232	7.0	8.0
164	179	10.0	11.4
165	310	6.5	9.8
171	319	8.0	9.5
174	258	7.3	10.2
192	365	5.9	8.9
204	325	9.6	10.4
207	242	7.8	8.8
334	422	10.9	10.8
363	410	8.6	8.5

PHOSPHORUS MOVEMENT IN CALCAREOUS SOILS IRRIGATED WITH WASTE WATER FROM POTATO PROCESSING PLANTS

Recent regulations governing waste water discharge into surface waters and increasing costs for waste water treatment have encouraged food processors to develop cropland irrigation systems for processing plant waste water disposal. To be effective, land application must remove the phosphorus from waste water before it enters surface or ground waters. Although considerable data are available for predicting and modeling orthophosphate movement in soils, the P in food processing plant wastes may be predominantly of plant or animal origin, and may be combined in organic phosphate forms.

The compounds do not have the same chemical and physical properties as orthophosphate and cannot be expected to react the same with the soil. Applying high-energy or high chemical oxygen demand (COD) materials found in food processing waste waters to soils stimulates biological activity and may solubilize otherwise insoluble soil P compounds. Under these conditions, P movement may be greater than predicted by orthophosphate models. Phosphorus movement depends upon total P application rates, relative amounts in organic forms, solubilities of the P forms, hydrolysis rates to orthophosphate, orthophosphate fixation, precipitation rates, water retention times, and water movement within the soil profile. Recent studies showed that the movement of surface-applied soluble organic P fertilizers in soils was directly related to the water infiltration rates and to the rates at which the applied material was hydrolyzed by soil phosphatase enzymes. This paper reports P movement under/ and application of potato processing plant waste water which is high in both COD and total P. Waste water samples from five southern Idaho potato processing plants and soil water extracts from waste disposal fields were sampled monthly for the first 3 years of disposal by land application.

Processing plant samples taken at 20- minute intervals over a 24-hour period with an automatic sampling device were composited into a container in a small freezer and immediately frozen and stored until analyzed. Soil water samples were extracted through porous ceramic cups cemented to the bottom of 38-mm-diameter polyvinylchloride tubes at depths of 0.15, 0.30, 0.60, 0.90, 1.20, and 1.50 m at two sites in disposal fields at each potato processing plant. Soil water was drawn into the tubes by applying 0.7 bar suction for 24 to 48 hours. After releasing the suction, samples were pumped from the sample tubes and transferred to storage bottles. Mercuric chloride was added to subsamples to retard hydrolysis of the organic and hydrolyzable P to orthophosphate. As soon as possible after bringing the samples into the laboratory, orthophosphate and hydrolyzable, organic, and total P concentrations were determined, Orthophosphate was measured by an ascorbic acid method without pretreatment of samples. Acidhydrolyzable P plus orthophosphate was determined by treating with a sulfuric-nitric acid mixture and autoclaving.

The combined phosphates were then measured by the ascorbic acid method. The orthophosphate concentration was subtracted from the value obtained to give acid-hydrolyzable P concentration. Total P was determined by converting all P in the samples to orthophosphate by autoclaving in the presence of potassium persulfate and sulfuric acid. Organic P values were obtained by subtracting the hydrolyzable plus orthophosphate from total P. Soil samples were taken annually at 0.3-m depth increments to 1.5•m and bicarbonate extractable and total P were de- termined.

Total P was determined by persulfate digestion. This method was compared with nitric-perchloric acid digestion and sodium carbonate fusion methods for total P before it was selected for use in this study, When the samples were autoclaved for 0.5 hours at 128C, the persulfate digestion gave total P values as high as the other methods.

The clay-size fraction of each soil depth increment was determined using a standard hydrometer method for particle-size analysis. This fraction was expressed as a depth of clay-size material by multiplying the clay content by the thickness of the depth increment. The increments for clay thickness (mm) were added to give a total depth of clay-size material for the 1.5-m depth of soil. Wastes from potato processing plants designated as IF, 2F, and 5F were disposed of by flood irrigation on nearly level borders. Those from plants 35 and 45 were sprinkled onto leveled land. Orchardgrass, alts fescue, or reed canary grass, or mixtures of these species, planted at each disposal site, were harvested for hay or grazed by livestock.

OTIMIZING SOIL MOISTURE AND IRRIGATION MANAGEMENT

Lack of adequate soil moisture is a limiting variable in sweet potato production. There have been many examples of nearly 100 percent yield loss for whole fields of sweet potatoes. For several years, the type of damage caused by drought stress was attributed only to herbicide applications. Recent research showed that by turning off supplemental irrigation during the critical storage root initiation stages at conditions approaching wilting point, we were able to induce similar damage symptoms – extensive lignification and misshapen roots. Optimal soil moisture for sweet potatoes grown in Louisiana is 15-20 percent volumetric water content at a 6-inch soil depth.

This number represents 50 percent of field capacity for soil moisture and needs to be calibrated for specific field conditions. Field capacity represents moisture held in soil after excess water has drained. Available soil moisture can be determined using a soil moisture probe. Soil moisture use and requirements are affected by many processes including transpiration and evaporation.

- *Wilting point:* The minimal point of soil moisture a plant requires not to wilt.

- *Field capacity:* The amount of soil moisture or water content held in soil after excess water has drained away and the rate of downward movement has materially decreased, which usually takes place within two to three days after rain or irrigation in permeable soils of uniform structure and texture.
- *Volumetric water content:* The volumetric water content in the soil represents the *Calibrating volumetric water content for your soil type.* fraction of the total volume of soil that is occupied by the water contained in the soil.

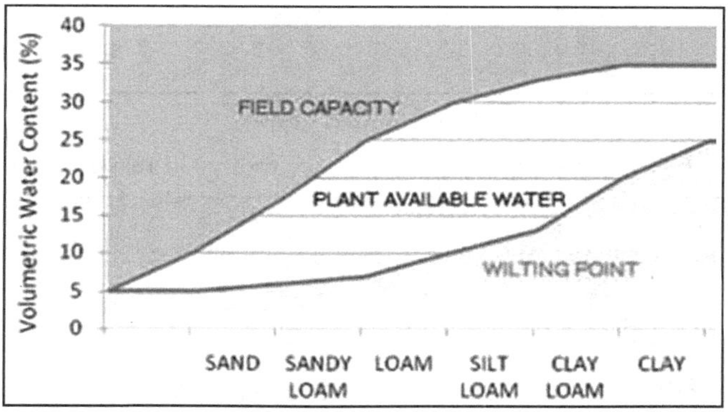

- *Lignification:* A physiological and developmental change in an adventitious root that typically yields nonproductive, pencil-type roots. Drought stress can result in lignification. Other biological and agriclimatic factors also may influence this process.
- *Storage root initiation:* The process whereby adventitious roots differentiate into sweet potato storage roots. This process occurs very early after transplanting and can be observed anatomically as early as 13 days after transplanting.

IRRIGATION MANAGEMENT

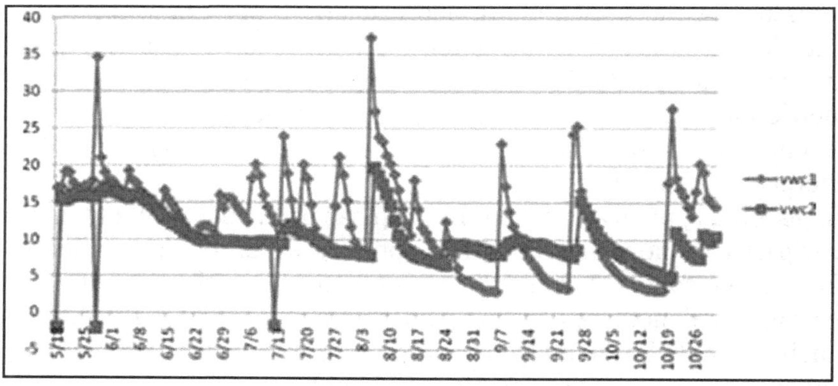

Irrigated graph

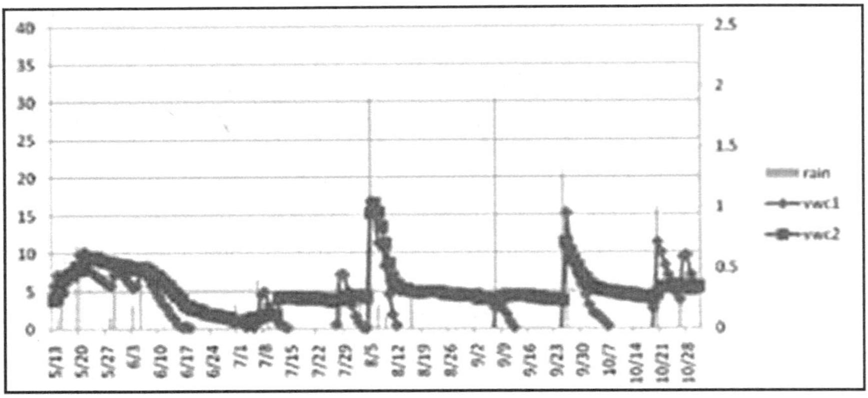

Nonirrigated graph

Fig. Soil moisture data for storage root yield responses.

Figure. Storage root development at 60 days in irrigated plots. The field plots were planted on May 12, 2010, and observed on July 1, 2010. The nonirrigated plots received about 1 inch of rain each month in May and June. Irrigated plots received overhead irrigation from a traveling irrigation gun when soil moisture at the 6-inch depth approached 12 percent volumetric water content. Soil moisture in nonirrigated plots averaged 10 percent volumetric water content within 20 days after planting but remained below 10 percent volumetric water content until sampling. The average volumetric water content at the 6-inch depth for irrigated plots was 15 percent during the first 40 days and averaged 10 percent thereafter.

In general, apply water when the crop has used about half of the available water capacity in the root zone. When applying water, don't completely fill or overfill the root zone. Overfilling leaches chemicals, such as nitrate/ nitrogen, wastes water and increases costs. Leave room in the soil for storing about 0.5 to 1 inch of rainfall that might occur soon after you irrigate. An irrigation management plan for sweet potatoes should be based on optimizing soil moisture during the critical storage root initiation period and for supporting maximum storage root bulking until harvest. This requires that moisture be present around the propagule at transplanting. Transplanting depth typically varies from 2 to 6 inches.

After five to 10 days, root depth can vary from 4 to 8 inches, depending on moisture uniformity and the presence of a hard pan. After 10 to 20 days, storage root length is determined, and the bulk of the root system is within the upper 12 to 15 inch depth of soil, depending on soil moisture uniformity and the presence of a hard pan, with lateral root growth accounting for nearly 80 percent of the root volume. The lateral root system accounts for at least 80 percent of the total soil water absorption. The lateral roots branch out at 85- to 90-degree angles, indicating that the root system will be able to efficiently use soil moisture at the 15- to 20-inch depth. Starting at about 30 days after planting until harvest time, lateral root lengths differ as a response to spatial and temporal soil moisture variations while the main storage root expands in diameter. In lighttextured soils or fields without a hardpan, overirrigation can lead to undesirable lengthening of storage roots in some varieties.

Fig. Center Pivot Irrigation of a Sweet Potato Crop

IRRIGATION METHODS

Center Pivot Irrigation

Center pivot irrigation is the most efficient means of providing supplemental water to a sweet potato crop. Timing and placement of the water are more accurately controlled compared to other irrigation systems. Center pivots can be effectively used during all stages of production. In addition, center pivots can be used during harvest to increase soil moisture and improve harvesting conditions.

Furrow Irrigation

If properly managed, furrow irrigation can be an efficient method for irrigating an established sweet potato crop. This method may not be the most efficient method when used to help establish the crop, however. During transplanting, moisture is needed around the base of the propagule, which typically is set at the top of the ridge. This requirement typically necessitates a longer application time from furrow irrigation to enable the wet zone to reach the top of the ridge. Depending on the length of the run, this results in saturated conditions in the upstream end. In addition, a rainfall event can further saturate the field during the critical period of storage root initiation.

Travelling Irrigation Gun

Reel guns can be used during all phases of sweet potato production. They commonly are used after the crop has become established to supplement water during the season. This method also is used to water sweet potato fields prior to planting and during harvest to improve harvesting conditions.

POTATO YIELD AND SOIL PHYSICAL PROPERTIES AS AFFECTED BY SUBSOILING

Agriculture practices today include the use of heavy machinery both to prepare soil for cultivation and during growth and harvest. Heavy machines cause high pressure on the soil which leads to compaction. High soil strengths are known to restrict the root development and prevent the root from growing beneath the plough pan. The consequence may be reduced root system and limited area from which the plant can extract water and nutrients. The result may be stressed plants and reduced growth. The ideal soil for potato production is deep, well-drained and loose.

The potato crop is sensitive to water stress and more sensitive to fluctuations in soil water than most other crops. It requires high water availability and minimum fluctuations between irrigations to produce high yield- and good quality. The sensitivity to drought is most often explained by the potato plants relatively shallow root system, and low root:shoot ratio, which limit its capacity to extract water from deep soil.

Sandy soils, which often are used in potato production, seem to be especially susceptible to subsoil compaction. Plant roots can normally penetrate soils with strength up to 2 to 3 MPa, but potato roots are more sensitive to high strengths. Already at a pressure of 1 MPa the root growth is negatively effected. Not only the roots, but also the tubers can be affected by high soil strengths. The soil compaction may physically restrict developing tubers and reduce both yield and quality. Further more tuber set might be forced further up in the ridge where soil temperature are higher and moisture is more variable and may be limiting.

Figure. a) Cross section of soil, showing how a subsoiler is working. b) A schematic drawing over the subsoiler.

Subsoiling is a way to loosen up the plough pan by deeper tillage. The operation includes vertical fixed blades that are "cutting" the soil, in the end of each blade an angled extension is lifting the soil some cm to break the soil compaction. If subsoiling is performed in between rows it is called inter-row subsoiling, the most suitable depth appears to be 38 to 46.0 cm. In less compacted soils chiseling might be satisfactory to use. It is conducted with flexible metal rods instead of fixed blades, and lack lifting extensions. But the purpose is the same, to remove soil compaction.

In general, subsoiling decrease soil strength and bulk density which means an increase of large pores. The effect is that the roots can penetrate further down in the soil which may reduce stress caused by inadequate water and nutrient supply.

Research in this field has been carried out in several different countries during the past 30 years. The purpose of this report is to make a review over these trials and try to answer the questions; how is subsoiling affecting the soil, the plant and the potato yields? Is subsoiling something that shall be recommended to Swedish potato growers? And if so, when is the best time?

SUBSOILING: EFFECTS ON SOIL

Penetration Resistance and Bulk Density

Penetration resistance and bulk density is often used to describe the properties of a soil. Enhanced soil compaction means that the force needed to penetrate the soil is increasing. This force can be measured with a metal stick, called a Penetrometer. The metal stick is pressed in to the soil with a certain force; a calibrator is recording the pressure which the soil is generating in MPa. Since the soil strength is affecting root penetration, and potato roots are particularly sensitive to high soil strength, this is essential to know for potato growers.

Bulk density is directly connected to the porosity and is a way to describe how compact a soil is. When the soil is compacted e.g. by the use of heavy equipment, the pores are decreasing in volume and the bulk density is increasing. During the process of turning the soil over with a plough a compacted section, referred to as the plough pan is formed. The plough pan can be found at different depth depending on how deep the standard tillage usually is conducted.

In the trials of Miller and Martins on a sandy soil, the plough pan with a strength of 2 MPa, was found at a depth of about 30.0 cm. Whereas in the standard tilled treatment of Bishop and Grimes, soil strength near 2 MPa were established at 20.0 cm and extended through a depth of 50.0 cm. They report a reduction of soil strength to 0.6 MPa through 55.0 cm of soil when using inter-row subsoiling. Buxton and Zalewski measured the penetrometer resistance 30.0 cm below the potato beds and found that it was significantly reduced in five of six fields after precision chiseling. Also Ibrahim and Miller found that, subsoiled plots had much lower soil strength than those not subsoiled except from the soil near the surface. But on the other hand Holmstrom and Carter conclude that subsoiling showed marginal success in loosening the soil below the plough depth.

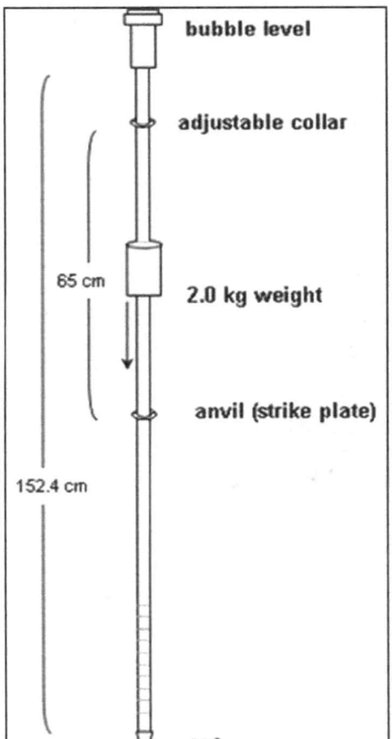

Fig. A Penetrometer

Pierce and Gaye Burpee tested two different tilling machines, Bush Hog Ro-till and Tye Paratill, when performing inter-row subsoiling. Both types of tillage decreased bulk density and soil strength, and increased the volume of large pores in the zone of compaction. The treatment lowered the soil strength down to 45.0 cm depth, and had an effect extending to a point midway between the potato beds. Measurements carried out by Pierce and Gaye Burpee showed that the bulk density in the soil depth interval 0.0 -7.6 cm and 10.0 -17.6 cm were not affected by subsoiling. But further down in the soil profile, at a depth from 23.0 to 30.6 cm, the bulk density were much higher in conventional tilled plots compared to the subsoiled ones. The trend in general was that zone tillage increased the amount of the larger pores, with radius classes over 36 ìm. Similar results were also presented by Sojka et al which found that the bulk density was unaffected by inter-row subsoiling within the top 15.0 cm, but reduced between 15.0 and 45.0 cm. However, Holmstrom and Carter reported a decreased bulk density in only five out of eight sites.

Infiltration rate

Soil compaction reduces the amount of pores, especially large pores. This is affecting the water holding capacity of the soil as well as the infiltration rate. Large pores are important when it comes to infiltration but also the moisture content in the soil. Analysis of the soil in the plough pan suggests that the reason for slowed down water movement is the shift from coarse pores to fine pores across the plough pan border. Since subsoiling increases the amount of large pores and lowers the bulk density it is likely to believe that the water infiltration rate would be positively affected by subsoiling. Haderson et al. also observed that less water were standing in the rows treated with inter-row subsoiling compared to conventionally tilled plots, but unfortunately no specific measurements were made to quantify this observation. Subsoiling may therefore be an effective way to decrease water erosion.

Although water infiltration rates were extremely slow in many commercial fields, there were no indication that slow infiltration rates were associated with low yield and quality. One explanation can be that the growers were able to overcome the unfavorable effects of slow water infiltration rates by proper irrigation management, including frequent irrigations for long periods. It seems reasonable that the increased infiltration rates which subsoiling gives would be beneficial during periods of excessive precipitation. But according to results reported from Soane et al. it seems to be the opposite, the yield is reduced in wet seasons when subsoiled.

Persistence and Recompaction

There is a gradually recompaction taking place in subsoiled fields. The persistence is dependent on several factors such as soil moisture at the time of subsoiling, the choice of subsoiling equipment and tillage- and traffic practices following the subsoiling. For e.g. recompaction caused by heavy machineries are less likely in drier soil. Soane et al. tested two different subsoiling equipments and found that both had a consistent effect upon the percentage of reductions in density and penetration resistance. However, subsoiling equipments which leaves an unbroken surface provides a protection to the loosened soil volume below against the effects of surface compacting loads. Therefore subsoiling equipments which disturb the upper most soil layer reduces its capacity to support surface loads, thus increasing the risk for recompaction. Holmstrom and Carter and Soane et al. found that trafficking, particularly mouldboard ploughing soon after subsoil loosening can cause significant recompaction of the loosened subsoil. Measurements done by Parker et al. three months after subsoiling showed reduced values of soil bulk density and penetration resistance. Further on, there was little evidence suggesting that any recompaction occurred during the following growing season. Pierce and Gaye Burpee found positive effects on soil physical properties immediately after zone tillage and the effect still remained in the second year after tilling. If compactions by machines are prevented by controlled wheel traffic the effects of subsoiling could last for three years;. Bishop and Grimes on the other hand has reported an intermediate level of soil strength already after one year of standard tillage operations and after two years of standard treatment the soil strength was almost back to the original level. Soane et al. found that the largest recompaction occurred between the second and third crop and the procedure went faster when root crops was grown in the fields.

Figure. 3. Diagram over soil strength showing five different treatments.

T_0= no subsoiling.

T_1= First one year of subsoiling followed by two years of conventional soil preparation.

T_2= First two year of subsoiling followed by one year of conventional soil preparation.

T_3= one year of subsoiling.

T_{123}= three years of subsoiling.

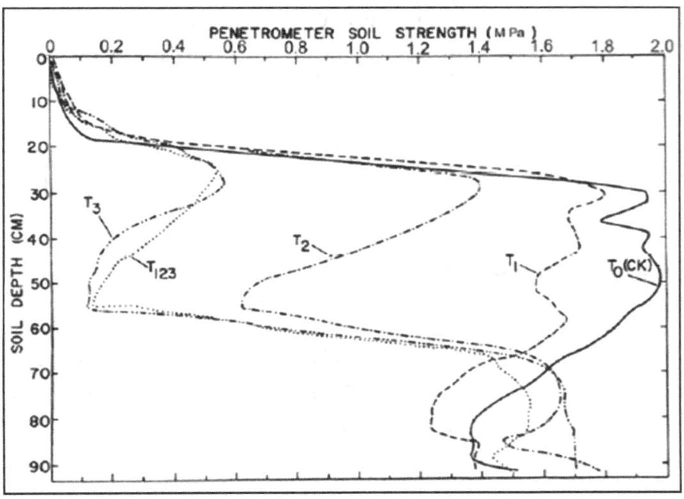

Negative Effects of Subsoiling

The result of subsoiling is closely connected to soil moisture content during and after subsoiling performance. It is well known that if the soil is wet machine operations can lead to an increase in soil compaction instead of loosen it. If precipitation occurs immediately after subsoiling the loosened soil might collapse. Subsoiling can cause other unexpected problems, which Holmstrom and Carter experienced. In five out of eight experimental fields, subsoiling resulted in an increased amount of surface stones, which were brought up from below. These caused extra work since they had to be removed before planting. Another problem that might occur when subsoiling is carried out in the spring or when the soil is too wet is the formation of clods; autumn tillage usually results in fewer soil clods at harvest than spring applied tillage practices. The clod formation is of concern because it requires extra labour at harvest to separate the clods from the potatoes.

SUBSOILING: EFFECTS ON PLANTS

Seed Tubers and Shoot Emergence

Inter-row subsoiling may improve the drainage, and when less water is occupying the pore space more air is kept in the soil. According to Sojka et al. and Westermann and Sojka that may have contribute to the higher soil temperatures of 0.5 °C observed in the depth of 5.0 to 15.0 cm of the subsoiled ridges. As subsoiling was done before planting the warmer and looser soil probably contributed to the earlier and more vigorous shoot emergence that was observed. The seed tubers on the subsoiled plots emerged three to four days before those in the conventional tilled plots. The earlier emergence observed resulted in a larger average dry weight of 45% measured about two weeks after emergence, at one of the six sites. Neither autumn tillage nor subsoiling seems to have any effect on the number of shoots per ridge;.

Subsoiling can be applied before or after seedbed preparation. Sojka et al. tried inter-row subsoiling after emergence and found that there were no substantial damage to the neither the seed tubers nor the sprouts. Although the soil-root contact seemed to be disturbed by inter-row subsoiling, the emergence appeared undisturbed. However these results seem to be valid only at an early stage of the plant growth. Halderson et al. found that delayed inter-row subsoiling gave a reduced plant size; looking at the period from emergence until full canopy, which probably was due to root disturbance. Other problems with late application were also detected in their trials. The height of the ridges was reduced by 1/3, and the tuber sprouts were exposed as much as 5.0 to 8.0 cm during inter-row subsoiling application in one of the experimental years. This led to a yield reduction especially of large tubers, but there were no increase in malformed tubers seen.

Roots and Water Uptake

By removing the plough pan the roots may penetrate further down in the soil than what they could have done otherwise. Bishop and Grimes and Ibrahim and Miller found a considerable increase in root density below 30.0 cm when the soil was loosened by subsoiling. With standard tillage, no roots extended below 61.0 cm, whereas inter-row subsoiling allowed significant rooting through 76.0 cm. The lower soil strengths also improved root growth vertical throughout the entire cross-section area of 20 and 40.0 cm measured from the bed center. The increased root penetration might not have any significant influence on the size of the total root system. The increased root penetration lead to an improvement in water uptake compared to no subsoiling, and by that the plants avoided water stress. A verification of the prevented water stress was increased leaf water potential, reduced stomata resistance and canopy temperature measured in those plants.

Connection between Subsoiling and Irrigation

Since potato crop is sensitive to water stress, an adequate supply of water is essential to get a strong, vital plant that will produce high yields. Sandy soils, which are known to have low water holding capacity, are commonly used in potato production. On these soils four days interval between irrigation is usually enough for visible water stress symptoms to develop and cause reduced yields. A comparison between daily irrigation and irrigation every fourth day showed that daily irrigation increased total tuber yield, number of tubers and specific gravity. But at the same time the tubers became smaller since the average tuber weight decreased. If a plant is stressed by the lack of water the stomata are closed and the photosynthesis slowed down. The plant produces less energy that can be stored in the tubers and in the end this leads to smaller yields.

If the irrigation interval of four days was applied but the plants were allowed to develop deeper root systems due to subsoiling, water stress were avoided. This led to an increase in yield and percentage of U.S No 1 (United States´ quality standards) tubers. The percentage of U.S No 1 tubers was low (46%) even with daily irrigation if no subsoiling were conducted and might be due to that the cultivar used, Russet Burbank is especially sensitive to water stress.

On loamy soils the results were similar to those on sandy soils. But since a loamy soil have a higher water holding capacity than sandy soil; the irrigation interval has to be longer to see any results. With weekly irrigation there was no significant outcome on tuber yield or grade due to subsoiling. If the period between irrigations were extended to two weeks this was enough for water stress to develop and both yield and percentage of U.S No 1 tubers were higher on the subsoiled than on the not subsoiled plots. Under near optimum irrigation levels subsoiling were of no benefit to the yield but during more drought conditions the additional rooting volume significantly increased potato yields, but had no effect on the quality. Therefore subsoiling can be of importance in areas were irrigation is not used. The extensive rooting can be essential for water uptake throughout the soil profile during periods of limited rain fall. When irrigation is applied frequently, the need for large amounts of stored water in the soil profile is not that crucial, and the subsoiling may not be needed.

Tuber quantity

The removed plough pan and the increased rooting might help the plant to avoid water stress. If this leads to improvements in tuber quality and quantity, it is important for the grower since it influence the economical outcome.

Bishop and Grimes found that the improved soil physical conditions, due to inter-row subsoiling, increased tuber yield with 6-10% and Pierce and Gaye Burpee reported generally increasing total yields but that it was only valid at seed tuber spacing of 25 and 28.0 cm, not at 36.0 cm. Henriksen et al. on the other hand saw no effect from inter-row subsoiling on the average total yield in two of three experimental years. The year when a significant increase in total yield was seen, the summer was very hot and dry. The connection between yield responses due to subsoiling and the climate and irrigation management has been investigated by Soane et al.. They found that yield benefits were obtained when all the following interaction factors occurred in combination: a droughty soil, a drought sensitive crop and a dry summer, and that the yield benefit was primarily due to the alleviation of moisture stress. When there was an absence of significant water stress or was there was no significant compaction in the unloosened subsoil nil response was obtained.

There are no benefits in applying excess water to the crop; this has been shown by Halderson et al.. They compared irrigation levels based on the evapotranspiration (ET) rate; the tested levels were 1.0, 1.2 and 1.4 times ET rate. The results showed that the total yields from inter-row subsoiled plots were equal to, or lower than the total yields from conventional tillage plots at all irrigation levels and tillage depth. A factor that might have contributed to the negative result was that the roots could have been disturbed during the inter-row subsoiling. A positive result from this study was that the yield of large tubers increased at optimal irrigation.

The effect on marketable yield was ambiguous, the first year it increased at both seed tuber distances whereas it decreased the next year at 36.0 cm and increased at 25.0 cm seed tuber spacing. In a summer with dry climate and high temperatures, the increase in yield by inter-row subsoiling was larger that other years, but there were no difference in marketable yield between the two treatments. Also in the experiment of Henriksen et al. the results are unclear, only in one of three trial years an increase in marketable tuber yield was seen, in the other two it was unaffected by subsoiling treatment.

In a trial where furrow irrigation was used, the total tuber yield did not differ in the first year between the subsoiled and non subsoiled plots, but in the second year the total yield increased significantly. In both these years the total percentage of U.S No 1 tubers increased with 4.6 % and 5.7 % respectively. In the first year the U.S No 1 graded tubers increased most in the size range from 114-284g were as next year the trend was that the tubers over 284g increased. Sojka et al. reported increased percentage of U.S. No. 1 potatoes weighing more than 284g in both their experimental years with sprinkler irrigation, due to inter-row subsoiling. No negative results have been reported as a consequence of inter-row subsoiling. Buxton and Zalewski announced that both tuber yield and quality were unaffected by the treatment even on soils with severe soil compaction. Holmstrom and Carter concluded that subsoiling prior to the potato planting improved neither the crop yield nor the quality. Halderson et al. declare that inter-row subsoiling, applied after seed piece setting, tend to reduce total yield, while the yield of U.S. no 1 tubers were unaffected. Ross also found that potato tuber quality, as expressed by the percentage of U.S No 1 grade were not significantly affected by subsoiling for all irrigation rates. Neither the mean number of tubers per ridge nor root weight recovered was affected by any tillage treatment.

Tuber quality

Specific gravity is a measurement of the density of a material that it is conducted in water. If a material has high density it also has high specific gravity. When it comes to tubers the specific gravity is of interest because it tells how much starch the tubers contain; high specific gravity on the tubers is wanted in the starch industry. Two tubers of the same size can have different specific gravity depending on its internal composition.

Inter-row subsoiling cause no significant changes in specific gravity. However there was an indication of slight beneficial effect of subsoiling on tuber specific gravity at irrigation rates below 400 mm but this was not statistically significant. Yields of undersized and oversized tubers were not affected by zone tillage. The yield of medium sized tubers was largely unaffected by zone tillage. Effects of inter-row subsoiling on malformed tuber yield were inconsistent. In one year there were no effect, but in the following seasons it reduced the percentage of malformed potatoes from 11.0 % to 8.3% and from 11.8 to 8.8% respectively. Halderson et al. found that inter-row subsoiling decreased the percentage of malformed tubers one year but increased it in the following year. Increased amount of malformation among the tubers weighing over 284g was also reported from Pierce and Gaye Burpee who also found an increase of hollow heart disease.

Differences Between Cultivars

Different cultivars have different genetic characters that might influence their responsiveness to subsoiling. Russet Burbank is the most frequently used cultivar in the studied trials, other cultivars might give other results.

Miller and Martin tested three cultivars, Russet Burbank, Nooksack and Lemhi, and found that there was a significant difference in how they responded to subsoiling. Nooksack and Lemhi often reacted similarly, while Russet Burbank was the one standing out. With subsoiling and no irrigation, both tuber yield and quality increased compared to no subsoiling, and the effect was largest in Russet Burbank. The best improvements from subsoiling were seen on the percentage of U.S No 1 tubers with the most pronounced effects occurring during a dry summer.

In another trial conducted by Miller and Martin the cultivars Russet Noekotah, Norgold Russet and HiLite Russet were used. They found that yield and the size of the tubers were decreasing most in HiLite Russet when subsoiling was not applied. Noekotah and Norgold yielded about the same but Noekotah had fewer tubers, higher percentage of U.S. No. 1´s and fewer undersized tubers.

Internal defects and specific gravity were generally unaffected by subsoiling, but there were some minor difference among the cultivars. HiLite had more vascular necrosis than the other two cultivars and that resulted in a lower percentage of tubers without internal defects. The specific gravity was about the same in all cultivars.

APPLICATION FOR OFFICIAL SOIL SAMPLING FOR POTATO CYST NEMATODE

- Growers wishing to apply for land to be soil sampled either for the growing of seed potatoes for marketing or to comply with the plant health import regulations of overseas countries should complete the form attached and not the general conditions below. Completed application forms for soil sampling should be returned to the Fera Senior Plant Health and Seeds Inspector in the area where the soil sampling is to be carried out.
- General Conditions under which applications are invited and accepted
  - Applications for soil sampling in order to enter any seed potato crops intended for classification in Fera's statutory scheme, should be made not later than 15 February in the year of planting.
  - If your application is for seed potatoes for marketing, please indicate for each field on page 2 of form PH 13 whether you wish to be able to market the seed as meeting the requirements for movement into a rhizomania Protected Zone.
  - If Zpd 1 is required you must tell us if the field is known to be infected with rhizomania.
  - If rhizomania is not known to be present in the field you must tell us if any types of beet have been grown in the field since 2003.

- Any certificate of the result of soil testing shall be used only in respect of any premises to which it purports to relate, and then only for the purpose for which the testing was carried out.
- No guarantee is given by Fera that any soil testing carried out will comply with or meet the requirements of any overseas country or importer.
- Fera and its officers shall be entitled to rely on the accuracy of all information supplied by or on behalf of an applicant for soil testing.
- The issue of any certificate as a result of soil sampling does not provide eligibility of a crop of seed potatoes for entry into a Fera classification scheme unless such a crop meets the rotational and other conditions for entry into the scheme.

- Applicants should be aware that under the Plant Health Order 1993 the future cropping of any land officially declared to be infested with Potato Cyst Nematode will be restricted as mentioned below.
  - Seed Potatoes: The growing of seed potatoes will be prohibited.
  - Ware Potatoes: The growing of ware potatoes will be prohibited except where the potato varieties to be grown are resistant to Potato Cyst Nematode or where Fera authorizes the growing of a crop to be harvested before the cysts mature or where a soil treatment approved by Fera for control of Potato Cyst Nematode is used.
  - Plants for Transplanting: The removal of plants for transplanting elsewhere will be prohibited, except where they are to be transplanted on the infested land.
- Where the applicant is not the owner of the land it is the applicants responsibility to ensure that:
  - The owner understands the cropping restrictions that will be placed on the land if Potato Cyst Nematode is found; and
  - They have the owners agreement for the sampling to go ahead.
- Fees will be charged for official soil sampling in connection with the export of potatoes, bulbs or plants. Details of the fees are given in a separate explanatory leaflet available on the Fera website at: www.defra.gov.uk/fera/plants/forms/ plantHealthForms.cfm or from your local Fera PHSI office. DON NOT SEND ANY MONEY WITH THIS FORM – an invoice will be sent to you later. Prompt settlement of the invoice will help to meet the overall cost of providing these services and may therefore affect the level of fees.
- Data Protection Act 1998: Information provided will be used for the processing of your application. It can also be used to determine the eligibility of your crops for export. Details will be held in our computer system and may be used to compile statistics on activity in your sector

and to inform Fera Plant Health and Seeds Inspectors, Defra, other Government Departments of their appointed agents of the location of crops in emergency situations, such as pest outbreak. It may also be used by Fera for scientific research. Fera may be required to release information, including personal data and commercial information, on request under the Environmental Information Regulations, the Code of Practice on Access to Government Information of the Freedom of Information Act 2000. However, Fera will not permit any unwarranted breach of confidentiality not will we act in contravention of our obligations under the Data Protection Act 1998. Fera, Defra or its appointed agents may use the name, address and other details on your application form to contact you in connection with occasional customer research aimed at improving the services that Fera provides to you.

IMPACT OF HIGH SOIL PH ON POTATO YIELDS AND GRADE LOSSES TO COMMON SCAB

Common scab caused by *Streptomyces scabies* Lambert and Loria is a problem on potatoes world wide, primarily due to its adverse effects on tuber appearance. Once established in a field, *S. scabies* can survive for extended periods on plant residues and consequently crop rotation provides only limited control of this pathogen. Seed, soil or foliarapplied chemicals provide only limited control of soil-borne scab. Fumigation may provide short-term control of soil-borne scab, but the economic and environmental costs of repeated fumigation are prohibitive.

Varietal resistance to scab is available but many of the varieties presently of commercial importance in North America have only limited scab resistance. The combination of these factors makes common scab one of the most costly diseases of potatoes in North America. The extent of infection and subsequent losses to common scab are influenced by soil moisture, nutrient levels, environmental conditions and crop management factors, such as planting and harvest dates. The common scab organism is capable of germinating and infecting its host over a wide range of soil pH, but the incidence and severity of scab damage to the tubers varies with pH. In areas with acidic soils, numerous studies have noted that both the number of tubers infected and the area of the tuber surface damaged tend to increase as the pH increases from 4.0 through 7.0.

Although liming these acid soils often increases yields, growers are cautioned against this practice as it increases the risk of scab. In the potato-producing regions of western Canada and the United States, soils commonly range from pH 7.0 to 8.5. Addition of sufficient amounts of acid-forming amendments to shift the pH of these highly buffered soils significantly below neutral is not practical.

An alternative approach to scab control on alkaline soils may be to lime the soil in an effort to increase the soil pH beyond the scab organism's tolerance range. Few studies have examined the impact of pH > 8.0 on the scab organism or the interactions between the pathogen and its host. Surveys of potato fields in New York showed that the incidence of scab was either unchanged or decreased as the soil pH increased above 7.0.

In a pot study, Blodgett and Cowan demonstrated that liming a soil from its original pH of 7.0 through to 9.0 decreased the incidence of scab from 80% down to 10%. The concept of using high soil pH levels to control common scab is complicated by the impact that high soil pH levels may have on the productivity of both the potato crop and subsequent crops in the rotation. This study evaluated the impact of altering the pH of an alkaline soil on crop yields, and the incidence and severity of common scab, with the objective of determining the soil pH level that produced maximum yields of marketable product.

The main trials were conducted from 1995 through 1998 on the University of Saskatchewan Plant Science Department Potato Research plots located in Saskatoon, SK. The site features a Sutherland series sandy loam soil, pH 7.8–8.1, EC < 1 dS m–1, with 4% organic matter. The site had been cropped to potatoes on a 3-yr rotation with dryland barley for the past 20 yr. Approximately 20 t ha–1 of cattle manure were applied to the plots every 3 yr following the potato crop. Common scab has been a consistent problem on the site.

The study was conducted in a strip-plot design with four replicates. A soil pH gradient was created within each plot ranging from pH 7.0 to 9.0 in 0.5-unit increments. Each individual plot was 4 m wide and 8 m long. The pH gradient range was generated 3 wk prior to planting by applying appropriate quantities of finely powdered calcitic limestone or elemental S to the soil surface, then incorporating to a depth of 30 cm with a rotovator. The amounts of CaCO3 or S required to generate each pH level were initially estimated from published liming curves and then finetuned in pot studies using soil from the test site. To achieve a pH of 7.5 required the application of 740 kg ha–1 of elemental S, with an additional 480 kg required to reach pH 7.0. Adding 1550 kg ha–1 of $CaCO_3$ shifted the pH up to 8.5, while an additional 1850 kg ha–1 of $CaCO_3$ was required to reach pH 9.0.

No attempt was made to balance the S and Ca added in the pH treatments as the test site had abundant soil reserves of these nutrients. Soil pH levels were tested 8 and 16 wk after treatment. Standard practices for production of irrigated potatoes in western Canada were used in all trials. Based on soil tests, sufficient N as 46-0-0 was broadcast and incorporated prior to planting to meet local recommendations for irrigated potatoes (residual + fertilizer N = 175 kg N ha–1). The crop was seeded in the second week of May using single cut or drop-type Elite 2 potatoes treated with the fungicide metiram as a seedpiece treatment. Metiram is not effective for the control of soil-borne common scab. The crop was seeded at 25-cm intervals in rows 1 m apart. Sufficient phosphorus was applied in a band adjacent to the seed to raise soil P2O5 levels to 120 kg ha–1. The cultivar used was Norland – an early maturing, red-skinned cultivar with moderate resistance to common scab. Weed control was achieved through pre-plant application of Eptam and Sencor followed by Gramoxone and Lorox at ground crack. A rolling cultivator was used after ground crack and again 2 wk later for inter-row tillage and to throw hills. Soil moisture levels were monitored in the plots using tensiometers and/or electrical resistance meters.

The plots were irrigated whenever soil water potentials fell below –60 kPa. No significant problems with insects or diseases were observed and the plots received no pesticides beyond those already described. The crop was desiccated using Reglone in the first week of September and harvested 2 wk later. The crop was initially graded based on tuber size and freedom from physical defects such as cracks. One hundred tubers randomly drawn from each treatment replicate were then individually evaluated to determine the percentage of each tuber's surface covered with scab lesions. According to the Canada Food Inspection Agency grading standards, potatoes meet No. 1 Grade standards if less than 5% of the tuber surface is affected by scab lesions.

Marketable yields in this study represent the tubers that met Grade No.1 standards for both size and freedom from excessive scab. Data for the 4 test years were initially pooled, as error variances for the soil pH, yield and scab data were homogeneous across years. Analysis of variance tests were conducted using the GLM program of SAS. The percentage data for the scab ratings were arcsinsquare root transformed prior to analysis. Years were treated as random effects in the model. As there were significant year ¥ pH level interactions for marketable yields and scab intensity the data for each year were then analyzed separately.

When a significant pH effect occurred on yield or scab levels, a profile analysis procedure was used to determine the threshold pH level where that effect became statistically significant. In this procedure, the value obtained at any pH level was contrasted with the mean values obtained from all lower pH levels. A second site was used to confirm the pH effects on yields.

This site features a Bradwell series sandy soil, EC = 1.0 dS m–1, 2% organic matter with a native pH of 8.1. In 1995–1997, test plots were established at pH levels of 7.5, 8.1, 8.5 and 9.0 using Norland potatoes. The trial was managed and data were collected as previously described. This site is free of common scab. The soil pH levels measured 8 wk after application of the amendments closely corresponded with the target levels. Later in the season there was greater variability in the pH values and the actual values observed in the limed treatments were consistently below the target value. Using very finely ground forms of both the S and lime appeared to allow the desired pH change early enough in the growing season to benefit the crop. However, finely ground materials are costly and create problems with product drift during application. Amending the soil in the fall prior to the potato cropping season would allow more time for coarser but more affordable forms of lime and S to work. The optimum pH range commonly reported for potatoes is 5.5 to 7.5, with higher pH levels causing significant yield reductions in some cases and little effect in others. The yield decline associated with high pH levels is usually attributed to pH-related deficiencies of P and Fe. In this trial, over a total of 7 site years at two test locations, soil pH ranging from 7.0 to 9.0 had no impact on yields prior to grade-out to scab.

No symptoms of P or Fe deficiency were apparent and the yields in all trials were well above regional norms. These results suggest that management practices that increase the soil pH to 9.0 pose relatively little threat to potato yields under the conditions and management practices common in western Canada and the United States. Blodgett and Cowan also noted little impact of increasing soil pH through to 9.0, but yields declined rapidly at higher pH levels. Scab levels and the scab responses to the pH treatments were consistent from 1995 through 1997, but in the 1998 trial there was little scab, irrespective of the soil pH level.

This likely reflects rotation of the trial in 1998 to an area of the field with low indigenous inoculum levels. Using elemental S to lower the soil pH from the initial level of 8.0 to 7.0 had no impact on scab levels. By contrast, in the three seasons in which scab was a problem, increasing the soil pH to 8.5 or preferably 9.0 significantly reduced both the incidence and severity of common scab.

The threshold soil pH values for scab control derived from this field study correspond with the thresholds observed in pot studies by Blodgett and Cowan. As increasing the soil pH reduced grade-out to scab without detrimentally affecting yields, marketable yields increased with soil pH.

Averaged over the 3 test years in which scab was a significant problem, yields after grade-out to excessive scab at soil pH 9.0 were 22% higher than at the native pH of 8.0. Even greater benefits might be expected with more-sensitive cultivars or where growing conditions and management practices were more conducive to scab.

How high soil pH suppresses common scab is unknown. In pure culture, high pH levels have no impact on germination or growth of scab. Conn and Lazarovits noted that the addition of poultry manure just prior to planting a potato crop provided a significant degree of scab control. Application of the manure temporarily increased the soil pH to above 8.0.

This pH shift was accompanied by an increase in levels of ammonia in the soil. Ammonia is known to be toxic to soil-borne pathogens such as *Fusarium oxysporum* and *Sclerotinia sclerotiorum* although its effects on *S. scabies* are unknown. This study demonstrates the potential for effective control of common scab by liming already alkaline soils. The requisite pH shift had no adverse impact on productivity of the potato crop; however, it may have detrimental effects on subsequent crops in the rotation. Similar crop tolerance problems occur when acid soils are treated with sulfur to shift the pH to below the tolerance range of the *S. scabies.*

The cost of the liming treatment will depend on the extent of pH change required, the buffering capacity of the soil and the cost of purchasing and applying the liming material. The benefits will depend on the extent of the scab problem, the degree of control achieved and the value of the crop. The material costs of this treatment may be reduced by limiting the lime application to the crop row. This is commonly done with phosphorus and certain pesticides. For any given field, a liming treatment may be required prior to each potato crop, as any effects of previous liming treatments may be lost in the three to four seasons between potato crops.

MONITORING OF MACRONUTRIENTS UPTAKE BY SOIL AND POTATO PLANTS – A COMPARATIVE STUDY

The macro and micro nutrients were effect the fertility of soil and growth of plants. The methods were described for determining Nitrogen, phosphorous & Potassium in plant tissues and soil, Its result indicate a correlation content with nutrients extraction and fertilizer applications. The aim of the paper is to determine the effect of macronutrient in fertility of soil and growth of potato plant. The concentration of nutrients in plant tissues can be measured in a plant extract obtained from fresh plant material,, as well as in whole dried plant material.

Total plant analysis is quantitative in nature and is more reliable and useful. Phosphorus, Potassium, Nitrogen is an essential element for all plants. Plants will grow slowly with low levels of Phosphorus, Potassium and Nitrogen in the soil; however, for agricultural production purposes, the soil should provide a sufficient concentration of P for optimum plant growth. While this goal is easily stated, it is difficult to determine whether or not soil P is sufficient for optimum plant production.

Consequently, much effort has been dedicated to the development of soil tests that determine the concentration of plant-available P, and whether or not that concentration is sufficient for optimum crop growth.Soil testing has been practiced in one. Nitrate in plant tissue are readily water soluble and is extracted by water from samples that have been oven dried and finely ground.

Nitrate is quantitatively reduced to nitrite. Soil test extract ants for phosphorus and potassium (K) were designed to rapidly assess the available nutrient status of soils. They serve as the basis for making recommendations for the addition of plant nutrients needed to achieve optimum yields. Most extract ants for P were developed to estimate the capacity of the soil to supply P to plants. They were designed to dissolve and/or desorb some fraction of the labile P and thus provide an index of the availability of P to crops over the growing season.

As the purpose of this study was to investigate the percentage composition of some important element, essential for growth and development of plants thus The resultant weight percentage of macronutrient present in potato plant and their respective soil sample is obtained from following methods all chemical used are „A grade, macronutrient were analyzed by different techniques (chemicals, reagent, flame photometry).

NITROGEN

5gm of sample was taken in kjeldahl flask, added 1 gm of CuSO4 and 5ml conc. H2SO4. Heated the mixture about 2 hour continue till the color of the mixture become more or less transparent.mixture was cooled and diluted with distilled water mixture was transfer to distillation unit.

Neutralized the solution mixture with conc.NaOH solution added a few glass beads. Taken 25ml of 0.1N.H2SO4 solution in a conical flask as a receiver. Heated the mixture, distillate was collect in the receiver, stopped heating and remove the flask. Distillate was titrate with.1N NaOHsolution, observed and total nitrogen calculated.

PHOSPHORUS

1 gm of dried powered sample was digested with 30 ml of 10% H2SO4 for half an hour on a water bath. The content filtrated and filtrate was transferred to beaker and 10 ml of conc. HNO3 were added to it. Mixture was warmed after stirring. The precipitate of ammonium phosphomolybdate so obtained was filtered through whatman filter paper no.44 and washed with dilute potassium nitrate solution. This precipitate was then dissolved in excess of N/10 standard NaOH solution and the excess alkali was titrated with N/10 standard HCl using phenolphthalein as indicator. Form the titer value, phosphorus content was determined.

POTASSIUM

Woodruffs method -5g of soil were taken in 100 ml screw capped bottle and 1.5ml of distilled water was added to it and was kept undisturbed overnight.35ml of 0.01m CaCl2 solution was added to it and contents were properly shaken for 1 hour and were filtered.K contents in the soil were determine using flame photometer which was calibrated with 0.01m CaCl2 solution.

Macronutrients in plant ash were analyzed by digestion followed by distillation of dry plants. N, P & K were analyzed by kjeldahl method, Spectrophotometric method & by flame photometer respectively

Table. Macro nutrients of soil samples of potato plant.

Industries	Macro Nutrients		
	Nitrogen (%)	Phosphorus(ppm)	Potassium (ppm)
R I I	4.88	60.2	148.3
S E M L	5.8	62.6	148
N I C O	4.6	61.9	148.2

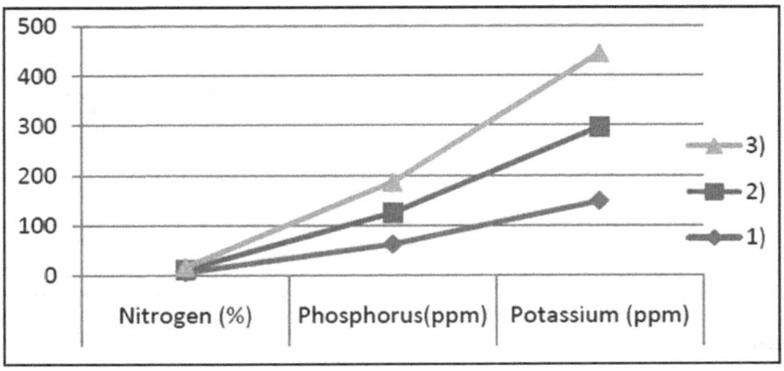

Fig. Graph macro nutrients of soil samples of potato plant

Table. Macro nutrient of samples potato plant ash.

S . no.	Macro Nutrients		
	Nitrogen (%)	Phosphorus	Potassium(ppm)
1)	17.45	64.23	103.3
2)	17.36	64.01	103.2
3)	17.48	64.22	103.3

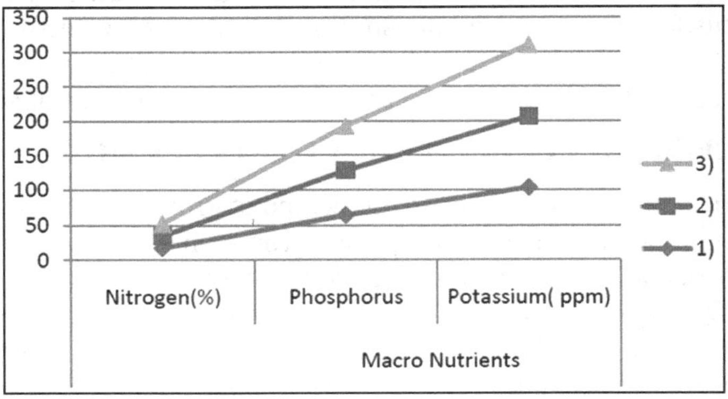

Fig. Graph macro nutrients of samples of potato plant ash

In collected soil samples of potato plants and their respective plant ash were analyzed by different method and procedure. After analyzing it was found that from macro nutrients the percentage of available nitrogen in soil sample was less than that of in their plant sample.

- The percentage of nitrogen (N) in soil of potato plant was obtained 5.6% and where as nitrogen percentage in plant ash was 17.45%.
- But the phosphorus and potassium are present in adequate amount in soil.As it was found that the concentration of phosphorus in part per million in soil of potato was 62ppm and 148.3ppm and in potato plant ash the concentration was 4.23ppm and 103.3ppm respectively.

Table. Comparative data of macro nutrients of soil samples of potato plant and plant ash.

Macronutrient	POTATO PLANT	
	Soil sample	Plant ash sample
N	5.6	17.45
P	62	64.23
K	148.2	103.3

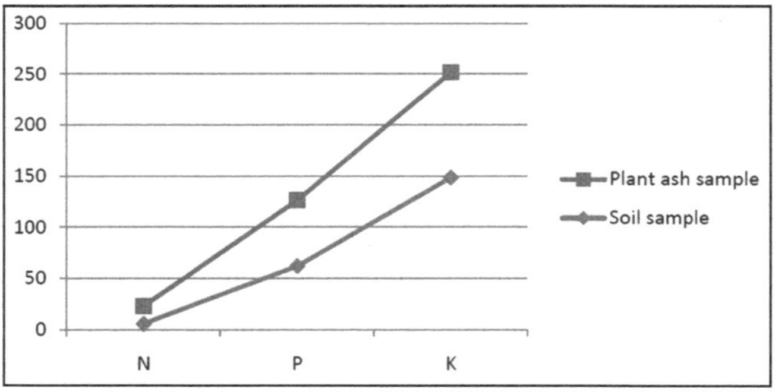

Fig. Macro nutrients of samples of soil of potato plant and potato plant ash

Considering the concept of "soil, plant nutrient index" the soil sample of their respective plant tomato and potato were found in category of moderate fertility status for potassium and phosphorus but having low fertility status for Nitrogen. Deficiency of nitrogen leads to sever disorder in crops such as narcosis. urea can be use to remediate the low fertility power of soil.

BENEFITS TO POTATO PRODUCTION & SOIL QUALITY

150,000 hectares of cropland are dedicated to growing potatoes across the UK. A large proportion of these crops are grown on light, sandy soils where crops are particularly vulnerable to reduced growth rates or failure during dry conditions. Quality compost is an excellent source of organic matter and using it will improve soil structure and increase the soil's water holding capacity, helping to protect crops and reduce irrigation requirements during dry spells. Heavier soils will also benefit as improved drainage can result in better germination and early root development.

WHAT IS QUALITY COMPOST

Quality compost is a natural product resulting from the controlled biological decomposition of biodegradable materials such as garden waste and waste food. The BSI PAS 100 compost certification scheme provides a baseline quality standard for compost, ensuring that it is consistently safe and reliable to use. The Quality Protocol for Compost provides a clear framework for the production and supply of quality compost. It builds on BSI PAS 100 and stipulates which waste materials can be used in quality compost production.

FERTILITY BUILDING

Quality compost provides crop available nutrients which can help to build up natural soil fertility and provide savings in the use of inorganic fertilisers. This is particularly true for potash, a key nutrient for potatoes.

Compost Type	Dry Matter %	Nitrogen (kg/t)		Other Nutrients (total, kg/t)			
		Total	Readily Available	Phosphate (P2O5)	Potash (K2O)	Sulphur (SO3)	Magnesium (MgO)
Green	60	7.5	<0.2	3.0	5.5	2.6	3.4
Green/Food	60	11	0.6	3.8	8.0	3.4	3.4

RECENT RESEARCH

WRAP has supported a number of research projects to investigate in more detail the benefits of using quality compost in potato production:

- Branston Potatoes have undertaken two projects in Perthshire, one on ware potatoes2 and one on salad potatoes3.
- In Herefordshire, ADAS investigated the use of quality compost on crisping potatoes.
- At two sites in East Anglia, a consortium led by Enviros investigated the use of quality compost on potato production as part of a long-term trial looking at improving soil health.

In addition to the field trials supported by WRAP, independent research into the benefits of compost has been funded by Defra through the 'Soil-QC' project and examined separately by ADAS through the 'ACORE' project. These projects have looked at consecutive compost applications over two to four years and provide valuable results on the longer term use of these materials in farming. Some of the results are used in this summary.

INCREASING SOIL ORGANIC MATTER

Compost additions are a valuable means of replenishing soil organic matter and can provide long-term benefits to soil structure and water holding capacity. Compost can also be used to improve drainage on heavy soils, opening up troublesome fields for healthy crop growth. Depending on total nitrogen content, green compost is typically applied at 30t/ha to comply with NVZ limits. This would supply 4-5 t/ha organic matter. Due to its higher organic content, a typical application of green/food compost would also supply about 5 t/ha of organic matter, although the total application rate might be only 20 t/ha.

The organic matter supplied by quality compost tends to be more long lasting than that in manures and other materials which are commonly applied to agricultural soils. This is because compost contains high levels of lignin which is a more stable, durable form of organic matter. As a result, repeat applications of quality compost will increase soil organic matter levels more quickly than manures and slurries.

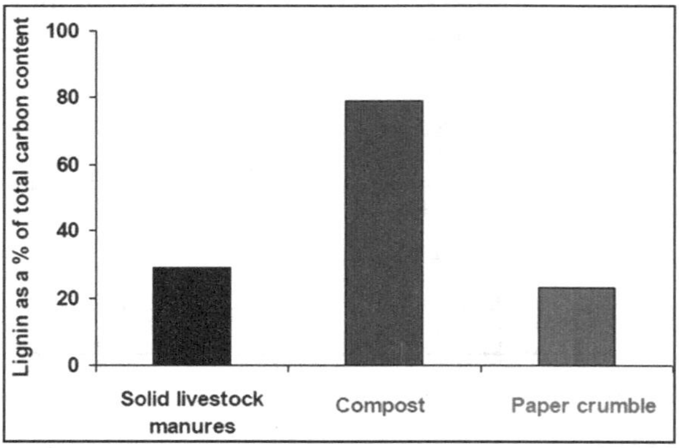

Fig. Lignin content of a range of organic materials applied to agricultural land.

Soil organic matter (SOM) has been measured in research projects on a variety of crops, including potatoes. Using quality compost over periods of between 2 and 10 years was found to increase SOM levels by an average of 12%, when compared with soils not receiving any organic matter.

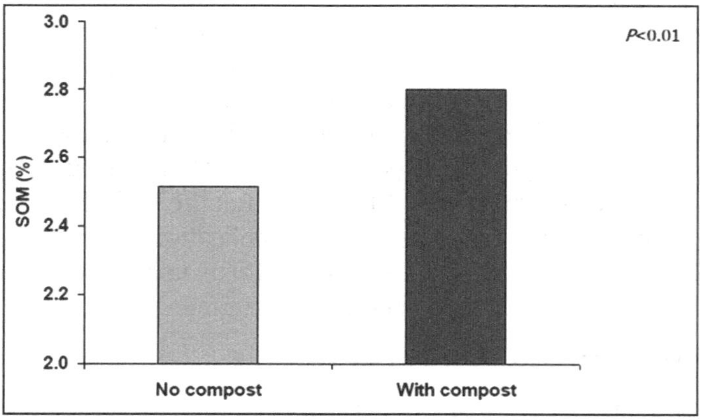

Fig. Effect of repeated compost additions on soil organic matter levels.

INCREASING PLANT AVAILABLE WATER SUPPLY

Increasing plant available water capacity (AWC) can have important benefits for crop yields, water use efficiency and farm profits, particularly on the sandy soils which are commonly used for growing potatoes and which have low AWC. Using quality compost improves the ability of crops to withstand periods of drought and decreases the need for irrigation.

AWC was measured across nine of the experimental sites and repeated applications of quality compost over 2-10 years were found to increase AWC by an average of 5%. The yield response of potatoes to applied irrigation water on sandy textured soils is approximately 0.25 t/ha/mm. A 5% increase in plant AWC is equivalent to an additional water supply of 3 mm in the top 30 cm of soil. For un-irrigated potatoes, this 'additional' water would result in 0.75 t/ha of extra yield.

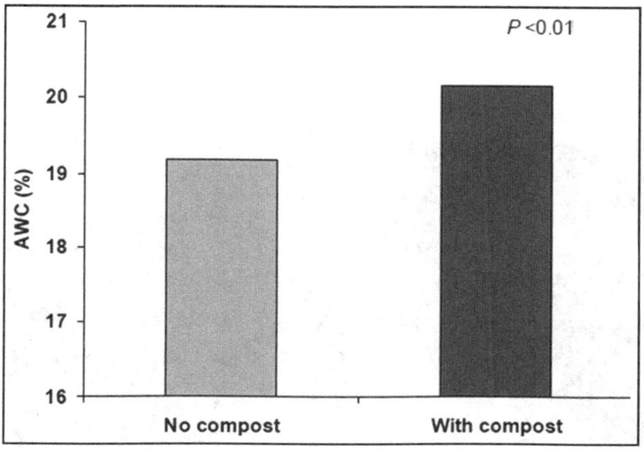

Fig. Effect of repeated compost additions on topsoil plant available water capacity (AWC)

COMMERCIAL IMPACT OF USING QUALITY COMPOST

Effects on Potato Yield

In all four WRAP projects, using quality compost either increased yields or had no impacts when compared with farmers standard treatments. In the Perthshire and Herefordshire trials, manufactured fertiliser applications were reduced to take account of the crop nutrients supplied by the compost, and potato yields remained comparable to the control plots which received full rates of manufactured fertiliser. In the Branston trial, using compost was found to reduce wastage and resulted in higher yields of saleable potatoes, although this result was not statistically significant over the short lifetime of the project.

Effects onPpotato Quality

Analysis from the WRAP trials showed that using both green and green/food quality compost has no negative impact on crop quality. In some cases, using compost was found to improve potato quality when compared with control plots of potatoes grown using conventional fertiliser practice. Notably, a reduction in common scab was found in one study and a reduction in waste potato tubers of c.4% was noted in another.

Under the BSI PAS 100, compost is tested for the presence of physical contaminants. Under the new 2011 version, compliant composts must contain less than 0.25% sharps. Even with this level of control in place, there are still some concerns that using compost may pose a risk of sharps contaminating potato crops and end products. In order to address this concern, the Branston project carried out industry recognised routine cook tests on sampled potatoes and demonstrated a negative result for sharps contamination.

Cost-benefit Analysis

Quality compost is a sustainable source of plant nutrients, particularly P and K, and its use can help reduce fertiliser bills. Three of the projects included a cost-benefit analysis which showed that, even taking into account purchase price, haulage and spreading costs, using compost in combination with a reduced rate of manufactured fertiliser rates can save £35-£110 per hectare.

6

Rotation

CROP ROTATION

Crop rotation moves different types of vegetable through the same garden bed over time. These are annual or biennial vegetables that go through their lifecycle, from seed to seed, within one or two years.

WHY ROTATE VEGETABLES?

- Avoid plant disease:
- Vegetables are rotated through a garden bed to reduce infection by plant diseases carried in the soil. Plants affected are usually those in the same botanical family, such as tomato and potato, which are both members of the *Solanaceae* family.
- Maintain plant health by reducing nutrient loss:
- Different plant types, such as leafy green vegetables or root crops, may make use of different nutrients in the soil. Continual cultivation of the same crop in the same place in the garden may, over time, result in nutrient deficiencies in the plants unless soil fer tility is maintained in other ways. It may help to grow tomatoes ever y 4-5 years, brassicas ever y 3-4 years and bean ever y 3 years, only once in the same garden bed.
- Crop rotation helps our plants to grow strong and healthy. Planting a different type of crop in the following growing season — such as a root crop (eg. potato) after a leafy green (eg. silverbeet) — reduces the likelihood of nutrient overuse and depletion.

A PROCESS OVER TIME

In crop rotation, different types of plants are alternated through the same garden bed over time:

- There are different types of crop rotation, such as:
- Rotation by plant type:
- *Legume* (peas and beans) then

- *Fruiting vegetable* (eg, capsicum) then
- *Leafy green vegetable* (eg. lettuce, cabbage) and,
- Finally, *root crop* (eg. potato)
- Rotation by plant family:
- Set up eight beds with these plant families:
- *Brassicaceae* (cabbage, cauliflower, broccoli, brussels sprouts, kohlrabi, radish, swede, turnip, rocket, mustard greens, mizuna, bok choy)
- *Solanaceae* (tomato, capsicum, potato, eggplant)
- *Fabacace* (broad bean, bean, peas)
- *Cucurbitaceae* (cucumber, squash, pumpkin, zucchini, watermelon, rockmelon)
- *Apiaceae* (carrot, celer y, coriander, dill, fennel, parsnip, parsley, mitsuba)
- *Chenopodiaceae* (silver beet, beetroot, spinach)
- *Asteraceae* (endive, lettuce, salisfy, globe ar tichoke, jerusalem ar tichoke, sunflower)
- *Amar yllidaceae* (onion, shallot, leek).

CREATING A GARDEN ROTATION PLAN

- *Step 1:* Divide your garden into beds or sections of similar size. The number of beds will depend on the length of the rotation plan you choose. A fouryear rotation needs four garden beds; a five-year rotation needs five beds, etc. Every year each garden bed is planted with a different group of plants. Planning the rotation carefully takes advantage of the conditions required and created by the different vegetables.
- *Step 2:* Place plants in groups. The suggested groups in this guide place plants based on family, nutrient needs, root depth, and growth season. There are also plant lists to help plan your own rotation; use them as a guide. Remember that the best gardeners are good observers. Pay attention to your garden, experiment, and even consider taking notes or keeping a garden journal to develop your own strategies.
- *Step 3:* Set up the rotation plan. In a four-year rotation plan, a garden bed that held group A plants the first year would hold group B plants the second year, group C plants the third year, and group D plants the fourth year. During the fifth year the bed once again holds group A plants and the cycle continues. Each garden bed starts with a different plant group during the first year, but follows the same order of plant rotations. In other words, A is always followed by B, which is followed by C, which is followed by D, which is followed by A. Tasks that help manage soil fertility such as adding slow-release fertilizer and soil amendments can also be included in your plan. For

example, potatoes can develop scab in soils that were recently limed, so lime is added after potatoes have been harvested. The soil fertility chores rotate with the plant groups.

CROPPING SEQUENCE AND ROTATION: IMPACT ON POTATO PRODUCTION AND SOIL CONDITION

PESTICIDE RESISTANCE

Pesticide resistance is defined as the ability of a pest population to survive a pesticide treatment to which the original population was susceptible. Weeds, disease pathogens, and insects are all pests capable of developing resistance. Pesticide-resistant organisms may occur naturally within a population in very low numbers. However, repeated use of pesticides with the same mode of action (MOA) (the sequence of events in which a pesticide kills a pest organism) allows the initially small percentage of resistant organisms in the population to survive and reproduce while susceptible organisms are killed.

As a result, the percentage of the pest population that is resistant to the MOA being used repeatedly becomes so large that the MOA is no longer effective on that pest population (Figure 1). There are several mechanisms by which a pest organism may be pesticide resistant. (1) The site within the organism where the pesticide acts (site of action) may be altered so that the pesticide can no longer interfere with the organism. (2) A resistant organism may metabolize or detoxify the pesticide more quickly than a susceptible organism; and (3) A resistant organism may be able to sequester the pesticide away from the site of action or block uptake or translocation so the pesticide never arrives at the site of action. Pest populations may develop multiple-resistance or cross-resistance. Multiple-resistance occurs when a population is resistant to pesticides with different MOAs. Populations with cross-resistance are resistant to different pesticides that all have the same MOA.

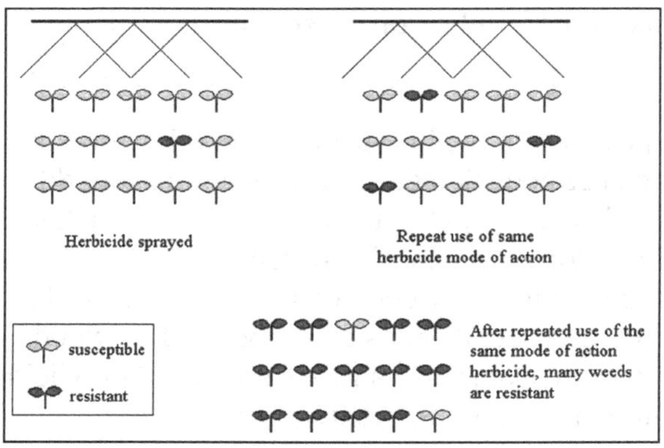

Fig. Resistant weed selection from repeated use of herbicides of the same mode of action. A similar effect occurs with disease pathogens, insects, and nematodes.

Development of pesticide resistance can be prevented or delayed by use of several good management strategies. One such strategy is alternating pesticides with different MOAs, both within and across seasons. Increasing the length of rotation also may help prevent the development of pesticide resistance by providing more chances to use MOAs different than what may be absolutely necessary to use in potatoes.

Herbicides such as EPTC (Eptam®), metribuzin (Sencor®), pendimethalin (Prowl®), rimsulfuron (Matrix®), or s-metolachlor (Dual Magnum®/Dual II Magnum®) are commonly and effectively used in potato fields in Idaho. Commonly used nematicides include Vapam®, Telone®, Mocap®, Temik®, and Vydate®. Strobilurin fungicides (e.g. Quadris®, Headline®) are commonly and effectively used in potato fields in Idaho for control of early blight. Increasing the frequency of potatoes grown on a particular field increases the likelihood of application of these pesticides and, as a result, increases the risk of development of pesticide resistance. The negative impacts of poor pesticide resistance management strategies can affect not only the fields in question, but resistant populations can spread to neighboring fields and even regions.

It is common knowledge that growing potatoes in the same location year after year is unwise for many reasons. One such reason is the likelihood of building up pest populations for which potatoes are a primary host and/or are more difficult to control in potatoes. Similarly, it is usually unwise to select rotational crops that do not provide at least one year of a break in a pest's life cycle.

Rotational crops often provide an opportunity to more effectively and/or inexpensively control certain pests. For example, inserting a grass crop (such as wheat, barley, or corn) between broadleaf crops (such as potatoes, sugar beets, and/or alfalfa) can allow for the use of a phenoxy herbicide, such as 2,4-D or 2,4-DB, which has the advantage of being less costly and having a different mode of action than most herbicides used in broadleaf crops. Other specific cropping sequence issues will be discussed in the sections below.

WEEDS

Ideally, the potato crop will be planted in soil from which the majority of the weed seed bank has been depleted. However, it is very difficult to completely eradicate weed seeds in the soil for a number of factors. Tillage can bring buried seeds to the germination zone near the surface. Depending on the weed species and environmental conditions, weed seeds can remain viable for long periods of time. In addition, new weed seeds are deposited by weeds not controlled in the previous cropping season, and also are transferred into fields by wind, water, birds, animals, humans, and equipment.

Cropping sequence choices may impact weed-related issues in potato production. Broadleaf weed control in grass crops is usually not as difficult as controlling broadleaf weeds in broadleaf crops. Multiple years of grass crops between years of potato production can be a cost effective means of reducing broadleaf weeds commonly found to be a problem in potatoes. As mentioned, the use of different MOAs from year to year is recommended to prevent or delay the development of herbicide-resistant weed populations. Since an increase in length of time before potatoes are again planted in a field usually increases the opportunities to use herbicides with different MOAs than what are used in potatoes, growers should take care to use these opportunites and choose herbicides for their rotational crops that have different MOAs than their potato herbicides.

For instance, a herbicide used in potatoes - rimsulfuron (Matrix®) has the same MOA as the following herbicides used in other crops: sugar beet - triflusulfuron (UpBeet®); small grains - imazamethabenz-methyl (Assert®) or thifensulfuron (Harmony GT®); corn – nicosulfuron (Accent®); and alfalfa - imazamox (Raptor®). All inhibit acetolactate synthase. The only way to ensure that herbicides with different modes of action are used through the cropping sequence is to understand the characteristics of each product that is used and know their MOAs.

Tank-mixing herbicides with different MOAs is another way to reduce the risk of developing herbicide-resistant weed populations. Rotating MOAs from season to season and tank-mixing herbicides with different MOAs can be considered a best management practice, even for fields with minor weed pressure. Weed control in fields with minor weed pressure usually can be accomplished with a single application of tank-mixed herbicides with differing MOAs and complementary control spectrum.

In addition to herbicide resistance issues, length of rotation can impact weed control and potato production costs in other ways. Short rotations can increase the occurrence and/or density of weed species that are hard to control in potatoes. Severe weed pressure, especially from weeds that are problematic in potatoes, can increase control costs and the incidence of weed escapes. Potato tuber quality and yield are reduced by weed competition and as a result, net returns are reduced. Fields with severe weed pressure may require an increase in the application rate(s) and/or the number of herbicide applications (i.e. an unplanned, post-emergence rescue treatment may be necessary if early-season weed control is not adequate).

Severe weed pressure also may require the addition of a third herbicide to an early-season, preemergence tank mixture in order to achieve effective control. University of Idaho research comparing the economic return of two- and three-way tank mixtures has shown that, in general, the increased potato tuber quality and yield advantage resulting from more consistent and possibly broader spectrum weed control with three-way tank mixtures than two-way tank mixtures overcomes the cost of adding that third herbicide to the mixture.

These tank mixture trials have been conducted for several years at the Aberdeen Research and Extension Center with moderate to severe mixed weed populations of hairy nightshade, redroot pigweed, common lambsquarters, kochia, green foxtail, and tame oat.

Including a green manure within the cropping system can also reduce weed populations and associated costs of control. Fall-planted rye, rapeseed, and Brassica green manure crops have all been shown to reduce weed populations in potato, although yields and economics are not always improved compared to standard weed control practices.

DISEASES

As far as diseases go, crop rotation is all about reducing inoculum. Some pathogens are so common in the soil that it is unlikely that any crop rotation will have a significant effect in reducing disease. Examples include:

- Bacterial soft rot (Erwinia carotovora)
- Fusarium dry rot (Fusarium sambucinum and F. coeruleum)

These diseases are typically managed through means other than crop rotation (good seed handling, seed treatments, fungicides, etc.).

Crop rotations can make it easier to manage some diseases. The pathogens causing the following diseases can survive in the soil, but adequate rotations (three years or more) will significantly reduce the soil as a source of inoculum. Seed-borne inoculum is a much more critical source of inoculum for these diseases in a good crop rotation.

The diseases in this category that are somewhat controlled with adequate rotations include:

- Rhizoctonia canker and black scurf (*Rhizoctonia solani)*
- Silver scurf (*Helminthosporium solani)*

The diseases for which soil-borne inoculum is most critical include (major soil-borne component):

- White mold (Sclerotinia sclerotiorum)
- Pink rot (Phytophthora erythroseptica)
- Pythium leak (various Pythium species)
- Verticillium wilt (Verticillium dahliae)
- Powdery scab (Spongospora subterranea subsp. subterranean)

As the time between susceptible hosts increases, the pathogen population decreases. Even though these pathogens can survive through three year rotations, increasing time between potato crops can lessen the impact of these diseases. Growing potatoes on a one or two year rotation will lead to greater problems with the diseases listed above compared to growing potatoes in a three plus year rotation. Because the organisms causing diseases with a major soil-borne component can live in the soil for several years, normal rotations (3-4 years) will not eliminate the pathogens, but will reduce the disease pressure in the field. As far as rotation crops as alternative hosts, dry beans and canola are susceptible to white mold. Growing these crops in rotation with potatoes can increase the severity of white mold. The organisms causing pink rot and Pythium leak can infect small grains. Research is currently underway to determine the importance of wheat and barley in the pink rot and Pythium leak disease cycles, and to see if other crops such as corn and sugarbeets are hosts. Verticillium has many alternative plant hosts, such as mint and many weed species. Fortunately, the main crops grown in rotation (small grains, sugarbeets, alfalfa, dry beans, and corn) with potatoes in Idaho are not alternative hosts for Verticillium. The powdery scab organism is not known to infect any of the other major agronomic crops grown in Idaho.

Incorporation of green manure crops has been highly effective in reducing disease pressure without necessarily reducing pathogen population numbers. This holds true even in short (two year) potato rotations. Green manure crops have different modes of action in reducing disease. Some green manure crops release volatile compounds that are similar to chemical fumigants when incorporated into soil. Others may do nothing to reduce pathogen populations, but instead increase soil organic matter, leading to an increase in microbial activity and competition that can suppress disease.

Overall, growing potatoes in a field more frequently than once every three to four years will increase the likelihood of soil-borne diseases. This will, in turn, put more pressure on other disease management practices. Sometimes this added pressure is enough to compromise the efficacy of other management practices and the long-term viability of pesticides due to developed resistance by the disease pathogen. Growers often produce potatoes on less than a three to four year rotation due to short-term economic decisions that are often faulty, but production costs are often lower for longer rotations and the long term cost of shortening rotations is increased disease pressure, higher chemical inputs, as well as increased likelihood of pesticide resistance.

INSECTS

Rotating crops is one of the most effective cultural control practices to reduce potato insect populations. Potato insects that over-winter in soil or on crop residue have increased populations as the frequency of potatoes increases in the rotation. Increasing the frequency of potatoes in the rotation also results in increased incidence of volunteer potatoes in rotational crops, acting as a host and launch site for flying insects, such as Colorado potato beetles. Planting cereal grains after potatoes aids in reducing migrations from overwintering sites to new fields. Colorado potato beetles are weak fliers and have difficulty flying out of foliage that is more than a few inches high.

Furthermore, the collective increase in the frequency of potatoes in rotations results in an increase in the number of acres of potatoes statewide and will impact the populations of insects invading from neighboring fields. There is a direct correlation with infestation incidence and distance of neighboring fields. Planting new potato fields as far from last year's fields as possible (at least 400 yards) will reduce the number of immigrant overwintering Colorado potato beetles into the new field. When good rotation practices are followed, beetles coming into a field will concentrate at the edge. In such a situation, spraying insecticides against this pest only on the infested edge may be adequate.

Cropping sequence also impacts insect pressure. Wireworms are becoming increasingly important in several states, including Idaho, and one of the possible reasons for this is the increased rotation with grasses for the cattle industry. Therefore, avoiding rotations that include clovers and grasses may reduce wireworm populations.

However, wireworms in general cannot be controlled with crop rotation itself because of the long life cycle of the most common species. Disturbing soils with cultivation (direct injury and exposure to desiccation or attack by predators), rotating crops, and controlling weeds are the best cultural control techniques available. Because soil dryness can kill many wireworms in an infested field, fallowing a field without irrigation will reduce wireworm numbers, but the control achieved must be weighed against the income lost from missing a crop year. Multiple fall plowing is known to be effective for the destruction of sugarbeet wireworm pupae, the most abundant wireworm in Idaho.

Planting alfalfa may reduce wireworm populations. However, leather jacket damage can be severe in potatoes following alfalfa, especially if the alfalfa is not plowed under until spring. Spider mites can be a problem following corn, bean, alfalfa, and clover seed if volunteer plant populations are high or if downwind from infested fields or along dusty roads.

As mentioned previously, a short rotation may increase the number of volunteer potatoes. Volunteer potatoes have been shown to be sources of both viruses and green peach aphid, which is the most efficient vector of potato viruses. Significant numbers of winged aphids can be produced in volunteer potato after plants have emerged. Aphids moving directly from volunteer potato plants in neighboring fields often transport viruses, since volunteer potato plants often have a high rate of disease infection.

NEMATODES

Nematode infestation on a potato crop results in yield decline and reduction in quality, thereby contributing economic loss to the industry. Among all nematodes, four species cause the majority of damage in Idaho potato fields, namely: root-knot, stubby-root, root-lesion, and potato-rot nematodes. Increased nematode pressure negatively impacts potato production in several ways. Root lesion and root-knot nematodes feed on roots and reduce yield. Root-knot nematodes feeding on tubers reduce quality by introducing physical imperfections at and near the tuber surface. Stubby-root nematodes transmit tobacco rattle virus (TRV) directly to forming potato tubers to cause corky ring spot symptoms. Virus infection blemishes tubers and renders them unmarketable. Damage from potato-rot nematodes is observed as lesions that begin to rot tubers in storage.

Although application of nematicides or biocontrols is often effective in reducing the effects of nematodes, these treatments result in an increased cost of production. There is also concern regarding the frequent use of these chemicals on the activity of beneficial soil microorganisms.

Crop rotation and cropping sequence impacts nematode populations and are important considerations for the management of potato nematodes. Increasing frequency of potatoes in a cropping system results in increased pressure from nematodes. Increasing the time between potato crops, particularly with rotation crops that are poor hosts, results in reduced risk of nematode damage. In general, potato tuber yields are higher in rotational sequences that begin with wheat or barley than in the sequences that begin with potato, sugar beet or onion.

However, small grain crops are a host for many nematode species. Infection damage from Columbia root-knot (*M.chitwoodi*) nematode tends to be worse in potatoes following a small grain crop as compared to corn or other poorly susceptible hosts. In contrast, damage from northern root-knot nematode (*M. hapla*) tends to be worse following alfalfa. Resistant alfalfa varieties, such as Archer II, AmeriStand 444NT, AmeriStand 403T, Robust T&N and Roccus reduce populations of root-knot nematodes. Stubby-root nematodes have a wide range of hosts, including cereal crops. Host range studies for potato-rot nematodes have indicated that potatoes and snap beans are good hosts, red clover and corn are intermediate hosts, alfalfa is a poor host, and oat is a non-host.

The use of non-host green manure crops could potentially improve rotations by reducing dependence on nematicides. The oil-seed radish and mustard green manure crops effectively reduce populations of sugarbeet cyst and root-knot nematodes. Experiments conducted at micro plot and field level confirmed that rape seed,, Humus and oil radish,, *Raphanus sativus* reduced Columbia root-knot nematode population and increased the potato tuber yield and quality under Idaho conditions.

Utilization of such resistant cultivars can reduce nematode survival, thereby reducing nematode damage and increasing yield potential. However, good weed control is imperative for these rotational/green manure crops to reduce nematode populations, as weeds are often hosts for nematodes.

In addition to reducing soil densities of nematode population, other benefits of using green manure crops include increased yields of subsequent potato crops, improved soil tilth and water holding capacity, reduced nitrogen leaching into groundwater, weed suppression, reduced soil erosion by wind and water, and potential suppression of other potato pests.

To effectively reduce nematode populations, green manure crops require at least eight weeks growth with soil temperatures above 60° F, although new varieties may offer equivalent control with less time required. Nematode control is enhanced with green manure establishment in a fertile, loose soil with adequate moisture and good weed and volunteer potato control. The key to nematode control is to plow down a large quantity of chopped green manure biomass (foliage needs to be green at plowdown). Discing 1 to 2 times, plowing, ripping, and harrowing are recommended.

SOIL

Cropping sequence and rotation impacts soil chemical, physical, and biological properties, such as nutrient cycling, erosion potential, compaction, organic matter, and biological diversity and activity.

Potatoes take up relatively high rates of nutrients and harvested tubers remove some of these nutrients in large quantities. Potassium is the nutrient of greatest removal, with 300-400 lb-K2O/acre removed for a 500 cwt/acre crop. This nutrient removal can have a depleting effect as the frequency of potato crops increases in the rotation.

Removal rates of other crops can also be high. Corn silage and, especially, alfalfa also have high potassium removal rates in comparison to other rotational crops, such as small grains and sugarbeets. Soils contain varying amounts of reserve nutrients found in and among the mineral structures.

In some cases, the soil contains thousands of pounds of various nutrients, but these reserves can be depleted over time if the amount removed is not replaced through fertilization and irrigation water. The speed at which the reserve nutrients enter the biologically active pool of nutrients also varies and can be inhibited as the number of high demand crops is increased in the rotation.

Nitrogen is removed at a rate of about 240 lb-N/acre by a 500 cwt/ac potato crop. Phosphorus is removed at about 80 lb-P2O5/acre. Calcium, magnesium, sulfur, and chloride are also nutrients with relatively high removal rates with rates ranging between 20 and 50 lb/acre.

icronutrients other than chloride (zinc, iron, manganese, copper, boron, molybdenum, and nickel) are removed at a rate of less than a pound per acre per year (some less than a hundredth of a pound per acre).

The levels of any of these nutrients in the soil can also be lowered over time if not replaced. Of course, there is nothing wrong with "mining" the soil if these nutrients exist in great excess of plant need, which is often the case. However, depleting the soil to critical levels can impact plant health over time. For example, a potato-wheat-wheat-wheat rotation removes approximately 260 lb-P2O5/acre and 450 lb-K2O/acre over the four year cycle. Potato growers commonly apply in excess of the removal rate for the phosphorus and this nutrient is generally not depleted in a potato cropping system. However, the potassium is much more likely to be depleted if the removal rate is not replaced. Other factors, such as mineral fixation, leaching, and decreasing mineral solubility complicate the availability of nutrients, which is one reason why soil testing needs to be the standard for fertilization rather than just basing it on removal rates. Shortening the rotation to a potato-wheat-potato-wheat cycle increases the removal to approximately 280 lb-P2O5/acre and 780 lb-K2O/acre over the four years, most likely requiring an increase in the applied rate of both nutrients.

In addition to nutritional effects, increasing the frequency of potatoes in rotation increases the erosion potential. Potatoes typically are clean-tilled both before and after the season. Most other crops are more conducive to minimum tillage practices that leave a higher percentage of erosion preventing residues on the soil surface. Fall tillage is common prior to potato planting in order to reduce disease inoculum and to hasten straw decomposition.

Furthermore, at the end of the season the potato plant leaves little residue on the soil surface and harvesting this tuber crop is the equivalent of a thorough tillage operation. Therefore, it is typical for the soil to be exposed to a high wind and water erosion risk from the fall prior to the spring after potato production.

As a result, erosion increases as the frequency of potatoes in the rotation increases. Inclusion of other low-residue crops, such as onions, sugarbeets and beans, has a similar effect on erosion rates, which is compounded if they are root crops for the reasons cited previously.

Erosion results in the loss of valuable topsoil that contains important physical, chemical, and biological properties that are not as prevalent in subsoil materials. Research shows that yield losses occur if the layer of topsoil is less than 12-16 inches. Basin and minimum tillage have been shown to work effectively in potato cropping systems to reduce water erosion. Furthermore, planting a green manure cover crop during the fall before and after potato production greatly reduces erosion and increases the organic matter content of the soil.

In addition to erosion benefits, organic matter is important for many reasons. Organic matter is a source of slow-release nutrients and the site of water and nutrient holding capacity. Organic matter also provides the "glue" that holds soil aggregates together, which is important for good soil structure and facilitation of vigorous root exploration and adequate water, air, and nutrient movement in the soil.

The primary component of organic matter is carbon, which is the source of energy for plants and microbes. As organic matter increases, microbial diversity and activity increases. The carbon dioxide released from respiring microbes is slowly released from the soil and captured by the canopy for use in photosynthetic processes. In addition, carbon dioxide is a greenhouse gas and there is a popularly perceived benefit if more carbon is stored in the soil and less is in the atmosphere.

The amount of organic matter being returned to the soil decreases as the number of low residue crops, such as potatoes, increases in the cropping system. High residue crops such as corn and small grains can add significant quantities of organic matter into the soil profile. Harvesting the wheat straw and harvesting corn as silage will reduce the amount of organic matter added to the soil, but these scenarios still result in more organic matter addition as compared to a potato crop.

In addition to residue issues, potatoes typically have high levels of tillage. Tillage warms and dries the soil, resulting in an increase in chemical oxidation and microbial decomposition of organic matter. The net effect of tillage is a significant loss of soil organic matter.

Tillage and organic matter loss also influence compaction. Compaction is a major source of yield reduction in many soils. Harvest and tillage operations in potato production result in destruction of soil aggregates by physical breakage and loss of organic matter binding.

Furthermore, field traffic, especially when wet, compacts the soil and results in less pore space for root, water, nutrient, and air movement. As mentioned previously, potatoes and other root crops have relatively more soil disturbance than most other crops. In addition, potatoes often require more trips across the field with heavy equipment.

The net effect of potato production is an increased risk of yield loss from compaction. Inclusion of other root crops, such as sugarbeets, into the rotation also increases compaction risk. Other crops, such as alfalfa, can have an ameliorating effect on soil structure.

Finally, soil biology is impacted by both frequency of potatoes in the rotation and the cropping sequence. Increasing the number of potato crops in the rotation favors the selection of potato pathogens, as discussed previously. Also mentioned previously, reducing the amount of residue in the cropping system reduces the food supply for microorganisms. Although microorganisms can be both good and bad for plants, the best scenario is to have large numbers of diverse species of microorganisms.

Repeated fumigation and low inputs of residues can decrease the total number microorganisms and the number of species present in the soil. The effects of fumigation on soil microbial populations seem to be most dramatic on soils with low base levels of organic matter.

Regardless of the soil type, microbial populations tend to fall dramatically immediately after fumigation, but then rebound with time. The make up of the specific microbial species in the population may change after fumigation. The long term effects of this type of disruption of the soil system are not well understood.

ECONOMICS

University of Idaho research has shown a 40 cwt/ac loss in yield for each year less than four between potato crops. Size and tuber quality have also been shown to decrease as the time between potato crops is reduced.And, as discussed above, increased pest pressure results in potential increases in herbicide, insecticide, fungicide, and fumigant applications. Relative rates of fertilizer may also increase as potato rotations are shortened. Obviously, increased costs with decreased yield and quality result in reduced net revenue for each potato acre and higher costs per hundredweight.

The long-term costs of short rotations are more difficult to quantify, but include loss of topsoil from erosion and increased likelihood of pesticide resistance. The cost of pesticide resistance is not only measured in terms of increased pest damage, but also results in an overall increase in pesticide costs as chemical companies have to develop new chemistries, while the cost recovery period for existing products is decreased, resulting in higher prices.

So, what is driving the trend to shorter rotations? There are many factors, but four tend to dominate. First, the potential gross return per acre (Yield x Price) is much higher for a potato crop than for grain or alfalfa, even when potato prices are only average.

And, while the chance of a super high potato price is rare, the "jackpot" is significant when it does happen. Second, once a grower invests in specialized potato equipment, no other crop offers the potential of recapturing the investment.

Potato equipment is expensive. Third, increasing the frequency of potatoes results in an increased economic efficiency of land (with fixed land payments). And fourth, potato equipment, like most farm equipment, has an economy of size, meaning the cost is less per acre for the larger equipment when used at an economically efficient level.

But the economic efficiency exists only when the costs are spread out over an optimum number of acres and the number of acres needed to achieve an economic optimum is greater for the larger-sized equipment.

For example, spreading out the cost of a new potato harvester over 500 acres rather than 400 acres results in a decrease in per acre ownership costs. But if the increased use starts to decrease the useful life of the machine, then the cost reduction will be less. The biggest perceived advantage to the grower is spreading the fixed payment on the harvester over more acres.

Long-term profitability is crucial to any sustainable rotation. One of the key management issues is balancing the increased economic efficiency for land and equipment (from shortening the rotation to increase potato acres and reduce machinery ownership costs) against the rising operating costs as potato frequency increases, as shown in Figure 2. Since shorter rotations often reduce yields and require more inputs, determining an accurate cost per hundredweight is essential to any grower struggling with this issue.

Each grower has to calculate these costs and balance them with the cost of production and revenue potential for the crops grown in rotation with potatoes.

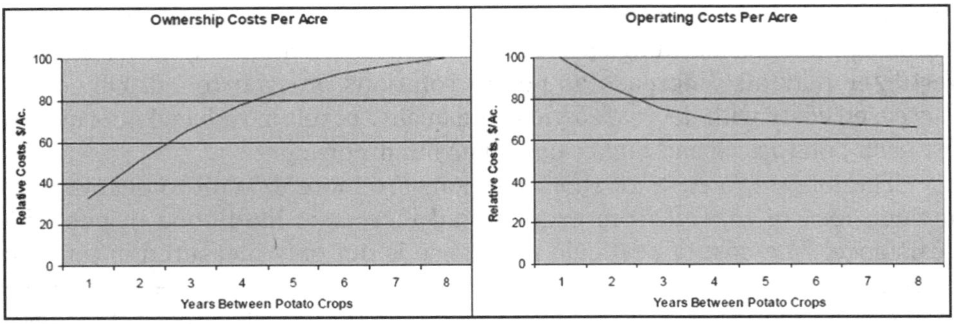

Fig. Estimated cost of production for both operating and ownership costs as a function of years between potato crops.

An example of such an exercise is given for a seed potato farm of 1350 acres in a short season climate (Table 1). The scenarios examined include four-, three-, and two-year rotations. The two-year rotation is evaluated both with and without fumigation.

These results show that the total cost per acre decreases as the rotation is shortened and potato acres increase, assuming a fixed farm size. The ownership costs per acre are lower because of an increased economic efficiency on machinery. But, the total cost per hundred weight (cwt.) increases as the rotation is shortened.

Lower yields on shorter rotations drives up the cost per hundredweight on both operating and ownership costs. Which is the best way to view this from a management perspective? The answer lies in the question "are growers selling potatoes on a per acre or a per cwt basis?," the later being the obvious answer.

Of course, this is only one example to illustrate this point. Other factors that need to be examined when evaluating the economics of entire cropping system are:

- What are the costs and returns to the rotational acre (not just the potato crop)?
- Is the land base fixed or flexible?
- How accurate are the yield projections?
- What is the accuracy of the cost estimates?
- What is the frequency of machinery replacement?
- Is the land charge fixed or variable?

Table. Cost of production example for a seed potato farm, comparing 2-, 3-, and 4-year rotations on a per acre and a per cwt. basis.

	4-Year	3-Year	2-Year	2-Year with Fumigation
Operating Cost/ac.	$1033	$1024	$1019	$1195
Ownership Cost/ac.	$496	$456	$402	$402
Total Cost/ac.	$1529	$1480	$1421	$1597
Operating Cost/cwt.	$4.13	$4.55	$5.10	$5.20
Ownership Cost/cwt	$1.98	$2.03	$2.01	$1.75
Total Cost/cwt.	$6.12	$6.58	$7.11	$6.94
Yield	250	225	200	230
Potato Acres	350	450	675	675

The land base is fixed at 1350 acres. Projections are based on a 25 cwt./ac. decrease in yield with each year less than four between potato crops (the research based average of 40 cwt./ac. is not used as the yield potential is relatively low for this field). Fumigation partially overcomes this yield loss. Operating costs increase as rotations are shortened based on increased pesticide costs. Some cost savings occur from reduced fertilizer as the yields decline. Ownership costs decrease as rotations are shortened based and potato machinery ownership costs being spread over more acres. In addition, the 2 year scenario has the addition of a second planter and replacement of the 2-row harvester with a 4-row harvester.

VOLUNTEER POTATO MANAGEMENT

Volunteer potatoes (*Solanum tuberosum*) are a prevalent weed in the Pacific Northwest (PNW) and other potato production areas of the world. Field surveys have shown that 25,600 to 186,155 tubers per acre are left in the soil following a commercial potato harvest. In areas where winter soil temperatures are not cold enough to kill tubers left in the field, serious weed problems occur. In the PNW, volunteer potatoes are a problem in cultivated crops such as alfalfa, beans, carrots, corn, onions, peas, potatoes, and wheat. They are especially damaging in non-competitive crops with few effective management options. In addition to being a competitive weed, volunteer potatoes harbor insect pests, nematodes, and disease, which partially negates the benefit of crop rotation. Volunteers may also contaminate potato seed production fields, rendering them useless as seed. While measures have been developed to help manage volunteer potato populations, no single method has been shown to be capable of reducing volunteer populations to noncompetitive levels. It is best to integrate many measures and use a holistic approach for volunteer potato management. This bulletin is intended to provide current information for volunteer potato management in the PNW. It will discuss the biology and ecology of the weed, its burden to production systems, and potential management strategies.

THE IMPORTANCE OF MANAGEMENT

Crop Competition

Volunteer potatoes are very competitive in rotational crops. They compete for light energy, nutrients, and water. In a study conducted at Paterson, Washington, volunteer potatoes reduced corn yield by 23 to 62 percent. Yield loss in onions and carrots has been estimated as high as 90 percent. Volunteer potatoes have also been shown to reduce yield in beans, sugar beets, and legumes. Yield losses also occur in more competitive crops such as mint, wheat, and alfalfa. Soil tuber density can increase over the course of the season if volunteer plants are not controlled (Figure 1). Volunteer potatoes complicate harvest operations and reduce the quality of several crops. Volunteer potato berries may be roughly the same color, size, and density of processed peas, which reduces their quality or adds to the expense of separating them.

Fig. This carrot crop will be unable to reach its full yield potential because of overwhelming competition

Diseases and Pests

To produce a high yielding, high quality crop, commercial potato producers must manage diseases, viruses, insects, and nematodes. Volunteer potatoes harbor diseases, insects, and nematodes that potentially increase the severity of these pests in future potato crops. Moreover, volunteer potatoes can serve as a primary source for disease and insects in adjacent fields, adding to the damage and management costs in commercial production fields. Volunteer potatoes can be a reservoir for late blight during the growing season. Late blight on volunteer potatoes may be untreatable due to the fact that effective fungicides are not labeled for use in the crop they infest. Research in the 1990s found volunteer potatoes to be the main overwintering host and the primary inoculum source for most potato viruses in Washington State.

Volunteer potatoes play a major role in the survival of potato leaf roll virus (PLRV) and potato virus Y (PVY). By hosting these viruses, volunteer potatoes enable them to survive, multiply, and be transmitted to commercial potato crops. Infected volunteer potatoes in rotation crops and commercial potato fields can be a catastrophic source of PLRV. The green peach aphid (*Myzus persicae*) is an insect pest and the primary vector of PLRV to susceptible varieties in seed and commercial production areas. Volunteer potatoes infected with PLRV in rotational crops are a source of inoculum in neighboring clean potato fields.

Green peach aphids will vector the virus from the infected volunteer potato to the neighboring commercial field, resulting in unacceptable levels of net necrosis in the tubers and possible rejection of the entire field by the processor. PLRV in seed potato fields may disqualify them from certification and therefore create unusable seed. Tubers generated from volunteer potatoes can host tobacco rattle virus, which is transmitted to potatoes by stubby root nematodes and causes corky ringspot disease. Other pests may use volunteer potatoes to persist through a crop rotation.

In Europe, plant parasitic nematodes such as the potato cyst nematode have been shown to multiply rather than decline in fields infested with volunteer potatoes. When growing a rotation crop free from volunteer potatoes, growers can experience a 35 percent reduction in nematode populations. Although seen as a potential benefit in reducing tuber numbers and vigor of volunteer potatoes, insects such as the Colorado potato beetle can build up tremendous populations on volunteer plants. Research in the Columbia Basin has shown that volunteer plants are the first food source to emerge in the spring and serve as the host for early emerging adult beetles. Although beetle defoliation of volunteers can be seen as a benefit, once the plants are defoliated, they may move on to nearby commercial fields and solanaceous weeds, allowing a more vigorous second generation to develop. The increase and persistence of disease, insects, and nematodes from the presence of volunteer potatoes are well documented and can be a detriment to the benefits of crop rotation. Thus, management costs will be returned not only in the rotational crop, but also in adjacent potato fields and when the infested field is returned to potatoes.

Contamination

Another problem with volunteer potatoes is the potential to contaminate rotation crops, which may severely reduce that crop's quality and value. Peas grown for canning or freezing must produce a continuous supply of clean raw product to the processor at harvest. Potato foliage and berries can contaminate the raw product and are very difficult to remove. Additionally, the presence of the toxic glycoalkaloids in volunteer potato foliage and berries may render an infested pea crop useless. Commercial seed potatoes are also affected by volunteer potatoes. Each state seed certification agency has its own quality standards for uniformity. Contamination by volunteers can result in mixed varieties being planted within the commercial crop. In addition to the problems with volunteer plants derived from tubers, volunteer potato populations can also be sustained and created in potato fields from true seed. In Europe, true seed potatoes persist for many years in potato fields.

BIOLOGY OF THE VOLUNTEER POTATO

Potatoes (*Solanum tuberosum*) are a perennial plant that can be established through vegetative propagation via tubers or from seed. Within U.S. production systems, volunteer potato plants almost exclusively originate from tubers rather than true seed.

Persistence in Soil and Overwinter Survival

In the PNW, potato harvest operations are the primary source of volunteer potato seed tubers in the soil. Tubers remaining in the field after harvest vary in size. In a survey of potato fields in the Columbia Basin the number of tuber leavings ranged from 7.6 to 126 cwt per acre (Newberry and Thornton). The majority of the tubers remaining in the soil were small, most weighing less than one ounce. Many factors were found to influence the size range and depth placement of the tubers, such as the use of preharvest vine desiccation.

Harvest from dead (desiccated) vines left fewer, smaller tubers in the field after harvest. Green (non-desiccated) vines tended to carry tubers of larger size and quantity off of the harvester, resulting in larger tubers remaining in the field at or near the soil surface (Newberry and Thornton). Following harvest, greater than 90 percent of remaining tubers are found within eight inches of the soil surface (Figure 2). In a Columbia Basin survey, 65 percent of tubers remaining after harvest were on the surface or in the top four inches of the soil (Newberry and Thornton). The remaining 35 percent located below four inches can total 25,742 tubers per acre, which is greater than a standard commercial planting (12,800 to 19,200 seed pieces/ acre). Because of the protective soil covering, these tubers are rarely exposed to temperatures that are low enough to render them non-viable. The large range of tuber depth placement means varying emergence times for volunteer potatoes in the spring.

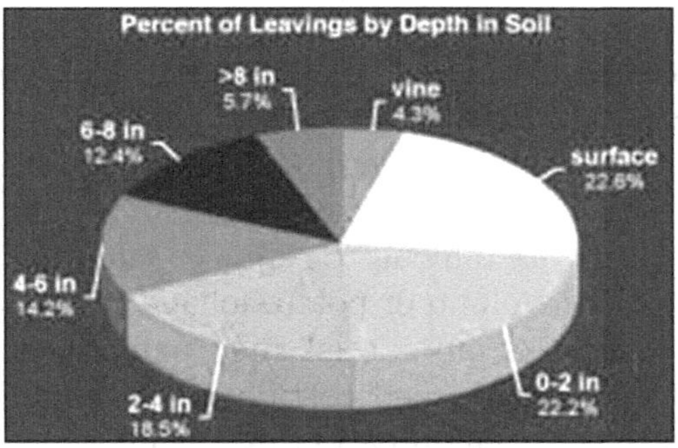

Fig. The percent of tubers remaining in the soil after harvest declines by depth in soil

Winter survival of volunteer potato is reduced by damage done from frost, birds, pests, disease, and physical harm from cultivation. In most of the PNW, temperatures do not remain cold enough for a sufficiently long period to kill all tubers left in the soil after harvest. In the Columbia Basin, tubers need to be exposed to temperatures of 28°F or lower to be rendered nonviable (Boydston and Seymour). A study from the Netherlands reported that tubers need to accumulate 50 frost hours below 28°F in order to be killed. Frost periods frequently kill tubers remaining in the upper layers of the soil in the PNW, but fail to affect more deeply buried tubers. In Othello, Washington, frost kills tubers below four inches only about one in four years (Thornton).

Even when frost can be counted on to kill tubers in the top four inches of the soil, approximately 19,123 tubers/acre still remain after harvest, a population high enough to cause great concern. Deep penetrating cold winter temperatures can result in complete control of volunteer potatoes with no effort or expense by the grower. In spite of occasional cold winter temperatures and rotation crop competition, volunteer potatoes have been documented to regenerate for as long as 20 years. The first year after potato production normally has the greatest amount of volunteer potatoes, with populations decreasing over consecutive years if effective control measures are implemented. It is unlikely that individual tubers persist longer than one growing season; tubers remaining in the field develop new daughter tubers each year.

For example, some red tubers were harvested from a field in Western Washington that was planted to a white variety. The individual red potatoes were the result of multiple regenerations perpetuated from a planting three years earlier. Crop competition contributes to the reduction in volunteer populations. However, even with competitive rotational crops and cold temperatures, additional means are necessary to effectively manage volunteer potatoes.

Plant Initiation and Emergence

Regeneration of growth from tubers occurs in early spring, once soil temperatures reach favorable levels. Emergence of volunteer plants typically starts in April and continues until the second week in June (Newberry and Thornton). The duration of the emergence period will vary from year to year because of annual changes in environmental conditions and variation in tuber size and depth of placement. Volunteer potatoes are able to generate vigorous regrowth due to the large carbohydrate reserve in the seed tuber or tuber piece. Individual tubers also have the ability to regenerate several sprouts because of multiple eyes, giving them the ability to survive multiple shoot removals. Vigorous plants have been produced from tubers no larger than garbanzo beans (approximately 3/8" in diameter) located as deep as eight inches in the soil (Figure 3).

Fig. Volunteer potato plant emerging from a small tuber buried over eight inches deep

Even the smallest tubers are able to survive the winter and produce a volunteer plant the following spring. Moreover, variable tuber size may contribute to variable emergence of the volunteers in the spring. This is important to keep in mind because various flushes of volunteers may impact management strategies.

Growth and Development

Once a tuber reinitiates growth, it develops very aggressively and competitively. Newly developing plants have been shown to live off energy reserves of the mother tuber for twenty or more days, giving them a competitive advantage over plants growing from a small seed. This advantage is even greater under stressful conditions. Although photosynthesis in volunteer potatoes begins as soon as the first leaf emerges, a significant amount of carbohydrates is still drawn from the mother tuber until new tuber initiation of the daughter plant.

Shoot removal tests have shown that volunteer potatoes have sufficient food reserves in the mother tuber to regrow new shoots following several removals of the initial shoots. This allows daughter tubers to form before initiating control measures, making management of volunteer potatoes more difficult in future years. By the time tuber initiation occurs, much of the food reserves in the mother tuber has been depleted, but re-sprouting from the mother tuber is still likely to occur.

As the volunteer potato grows, it depletes nutrients and water intended for the rotational crop. In less competitive crops, such as onions and carrots, the volunteer plants can be very effective at simply shading out the developing crop. The size of the mother tuber affects the competitiveness of the volunteer plant and the resulting size and number of daughter tubers. As the size of the mother tuber increases, stem number increases, resulting in a denser stand of volunteer potato plants and a rise in daughter tuber density. Therefore, plants developing from smaller mother tubers may be less competitive, easier to manage, and leave behind fewer daughter tubers than those originating from large mother tubers. Ultimately, the number and size of tubers produced by volunteer potatoes vary, depending on management methods, vigor, disease incidence, size of mother tubers, depth of burial, and crop competition.

Seed and Tuber Production

Volunteer potato plants have the ability to produce a large number of tubers as well as true seed. Potato true seed production ranges from 27 to 100 million seeds per acre in parts of the world that raise varieties capable of producing a true seed. While true seed production can be immense, the majority of the emerging volunteers in the PNW originate from daughter tubers, as most varieties grown in the area are not capable of producing a true seed. However, increased use of European cultivars in the United States may increase the potential for volunteer plants developing from true seed.

In the Columbia Basin, the number of tubers left after harvest ranged from 25,600 to 183,485 tubers per acre with an average of 73,550 tubers per acre (Newberry and Thornton). This is greater than four times the number of tubers used in commercial plantings (12,800 to 19,200 seed pieces/acre). The same study indicated the number of tubers with spring regrowth following a mild winter ranged from 1,600 to 41,000 plants per acre, with an average of 16,606 plants per acre. The study also showed significant reduction in spring regrowth following a winter in which damaging temperatures occurred. In the coldest year of the study, ten fields had complete kill of volunteer tubers and no volunteer plants were detected. The choice of rotational crop also has the ability to limit tuber number and size.

MANAGEMENT OPTIONS

Volunteer potatoes can be competitive as a weed, harbor harmful insects, and cause disease and nematode infestations. Moreover, they can regenerate within rotational crops to contaminate potato seed crops or commercial crops of potatoes. Management of volunteer potatoes is very important, extremely difficult, and can only be achieved by using integrated management methods. Volunteer potato management should consider three main goals: (1) reduce volunteer potato competition within rotational crops to prevent yield loss; (2) eliminate new tuber production by volunteer potatoes within rotational crops to reduce losses and control measures required in the succeeding crop; and (3) restrict the number of volunteer potatoes serving as hosts for potato diseases and pests. Throughout the remainder of this bulletin we will discuss potential management options within rotational crops. Each option will be evaluated based on the criteria established by these goals. Management options are broken down into five approaches: cultural, mechanical, biological, preventative, and chemical. Cultural management includes cropping practices that reduce the rate at which pests colonize a field or make the crop environment less suitable for pest survival. For volunteer potatoes, competitive crops may be chosen for a rotation and sound agronomic practices applied in the commercial potato crop to reduce the number of undersized tubers left in the field. Mechanical management involves the use of equipment to remove or destroy tubers and plants. An example of mechanical management would be a shallow tillage operation after harvest. Biological management utilizes living organisms or natural enemies to suppress populations of pests. Preventative management focuses on the cause of the problem, preventing the introduction of volunteer tubers to fields. Chemical management is one of the most commonly used methods for managing volunteer potatoes and may involve the use of herbicides and fumigants in rotational crops as well as maleic hydrazide, a sprout inhibitor, in the potato crop. The use of all of these management strategies must be combined in a collaborative holistic approach to address volunteer potato management.

Cultural

Crop rotation is practiced on most of the potato growing acres of the Pacific Northwest. Rotation presents an opportunity to choose competitive crops with cultivation and herbicide options to reduce the number and impact of volunteer potato plants. Winter or spring wheat or spring barley following potatoes provides strong competition to volunteer plants, compared to less vigorous shorter season crops like beans or peas. In the PNW, cereal grain competition reduced shoot emergence from tuber leavings and subsequent daughter tuber production by 50 percent, compared to a less competitive crop. Volunteer plants produced four daughter tubers in a less competitive open canopy crop, compared to only two daughter tubers per plant in a winter wheat crop.

However, herbicide or cultivation options for volunteer potato management in winter wheat are limited. Planting competitive, long season crops does not eliminate volunteer potatoes but it does reduce tuber number and size in comparison to growing a less competitive crop in rotation. Although less competitive crops may not be as effective at reducing the volunteer potato problem as those that are more competitive, any degree of crop competition is very important in the overall management strategy. Moreover, rotational crops, whether competitive or not, may provide effective chemical or mechanical tools to reduce volunteer potatoes.

Management of the commercial potato crop is an integral part of reducing the number of volunteer potatoes. A healthy potato crop with a uniform set of large, easily harvested tubers is less likely to have small tubers that fall through the chains on the potato harvester. Good potato health management practices involve proper seedbed preparation and spacing at planting, fertility management, soil moisture management, and reduction of diseases such as *Rhizoctonia* that, when present, reduce tuber size. A healthy commercial potato crop will not only pay for itself—it can also help reduce costs of managing volunteer potatoes. Grazing is another practice that has shown positive results in reducing volunteer tuber numbers. In Tasmania, sheep are allowed to graze potato fields after harvest, feeding on the tubers left in the fields. Cattle may also be effective, but are more prone to choke on the potatoes.

Mechanical

Mechanical tillage to manage volunteer potatoes can be used in conjunction with other management practices to reduce or suppress volunteer potato plant numbers. Washington studies have shown that light tillage in the fall reduces the number of emerging plants in the spring. Any type of deep cultivation that places tubers deeper in the soil where they can be protected from killing frosts should be avoided. A Washington study showed that it is not beneficial to use a moldboard plow for tillage after potato harvest.

The best options for tillage would be a rod cultivator or a paraplow. Another effective option is to fall plow following freezing temperatures to expose deeper tubers to additional winter cold temperatures. The greater the number of tubers left at the soil surface the better. Studies have shown that cultivation of volunteer potato plants in corn reduces daughter tuber numbers. In a Washington study, cultivating four times during the season greatly reduced volunteer plant vigor and subsequent daughter tuber number. Plants were cultivated when volunteers were at the six- to eight-leaf stage and stolons were hooking, followed by additional cultivations each time the plants re-grew to this stage.

Disking early emerging potato plants before planting corn also reduced early season competition in conventionally tilled corn. Preplant applications of glyphosate in no-till corn had the same effect. Herbicide treatments alone do not completely eliminate volunteer potatoes, however. Cultivating seven to ten days after a post-emergence herbicide application improves volunteer potato management and greatly reduces the number of daughter tubers produced. Another management option for reducing volunteer potatoes is to collect and crush potatoes on the harvester.

This is not a general practice, but has been considered in Scotland and may become more practical as technology advances. Harvesters can be designed with a method of collecting small tubers that fall through the chains. These are then sent through a crusher at the rear of the machine.

This method is compromised by small stones and may not be effective if tubers are too small. Another option is to use smaller pitch chains on the harvester to collect smaller tubers during a second harvest, and utilize them for special processing needs or animal feed.

Biological

Biological management with Colorado potato beetle (*Leptinotarsa decemlineata Say*) fits very well into an integrated volunteer potato management plan. Colorado potato beetle (CPB) is an insect that feeds on potato foliage. CPB can completely defoliate plants in a short amount of time, consequently reducing daughter tuber size and number. However, defoliation of volunteer potato plants by natural populations of CPB often occurs too late to prevent yield loss of the rotation crop or to substantially reduce new daughter tubers. Volunteer plant defoliation by CPB, when coupled with effective early season herbicides or other methods, can result in improved volunteer potato plant management and reduce the weight and size of daughter tubers produced.

Shoots emerging from smaller size tubers are less vigorous and therefore may be easier to manage in subsequent crop rotations. Insecticide use in rotational crops such as corn, onions, and carrots may need to be modified in order to allow CPB to defoliate volunteer potato plants and other solanaceous weeds. The use of early season herbicides in combination with later season feeding of the CPB has been shown to suppress daughter tuber production. Washington studies have shown a decrease in volunteer potato competition in corn and onions due to late season CPB defoliation. However, encouraging CPB populations on volunteer potatoes has the potential to be detrimental if they move to commercial potato crops after defoliating volunteer potatoes. Managers must evaluate the benefits versus the potential risks. In areas using insecticides at planting time or seed treatment insecticides, the CPB threat for commercial fields may not be a concern.

Preventative

Preventative methods can be highly effective and, next to killing low temperatures, can be one of the most economical strategies to implement. Preventative management strategies that can be integrated into a holistic approach include harvester management, harvesting after plants are dead rather than still green, and the application of a sprout inhibitor. In the Cultural Practice section of this bulletin, overall agronomic health of the commercial potato crop was discussed and should also be considered as a preventative management option. Good agronomic health of the potato crop pays dividends for the current crop and helps suppress volunteer potato numbers.

Harvest Management Considerations

Sound harvest practices can help producers get the most out of their current crop while reducing the tuber loss that leads to a volunteer potato problem in a rotation crop. Although the main concern during commercial potato harvest is bruise management, the following additional steps will help minimize tubers lost during harvest.

- Blade depth Manage blade depth to assure that all tubers are being removed from the soil. If the blade is too shallow, tubers will be sliced and a portion of individual tubers will be left in the soil.
- Haulm separation Operate harvester to achieve removal of tubers from stems so they do not get carried off of the harvester.
- Truck loading Position trucks receiving harvested tubers to prevent spillage.
- Tuber intake Operate harvesters to avoid pushing tubers out around the throat of the harvester.
- Forward speed Use the optimal ratio of forward speed to chain speed to maximize soil separation and tuber transport.
- Primary chain gap size Gaps between links on the primary chain should be properly sized to minimize the number of tubers that fall through. Of course, this must be compatible with intended market.

A factor to consider before harvest is the condition of the potato vines. Both premature vine senescence and green versus dead plants at harvest have been shown to affect tuber depth in the soil profile and the amount of tubers left after harvest. Many manageable agronomic factors contribute to premature vine senescence in the potato crop. These factors include fertility, soil moisture management, and disease and pest management. In a Washington study, fields that senesced early had a higher percentage of smaller tubers and consequently a greater number of tuber leavings than late senescing fields.

In fields where harvested plants were green, the average number and weight of tubers were more than double that of fields harvested when plants were dead. Tuber depth was also affected by plant senescence. Fields harvested green had 75 percent of the tuber leavings on or within four inches of the soil, compared to 34.2 percent for fields harvested when plants were dead.

When green plants are harvested, tubers are concentrated in the upper levels of the soil where they will more likely be subjected to freezing temperatures. Green plant harvest also leaves larger tubers in the field after harvest. Harvesting when plants are dead will not leave as many tubers in the upper portion of the soil and will not leave as many large tubers. Application of the sprout inhibitor maleic hydrazide (MH-30) to plants during the growing season can be used to suppress volunteer potato plant numbers. MH-30 is a growth regulator that supresses tuber sprouting when it is applied three to four weeks before normal harvest but at least two weeks before vine kill. The effectiveness of sprout suppression varies by cultivar and by tuber size. Studies found the residue level of MH-30 present in the small tubers that make up the majority of harvest leavings to be lower than that found in larger tubers normally taken to storage. In some cultivars, an application of MH-30 can significantly reduce the number of volunteer potato plants the following spring, especially if there are many large tubers in the leavings. Because of the variability in residual

MH-30 concentration in tubers of different cultivars and of different size, research is underway to determine optimal timing of applications. No conclusive results are currently available.

Chemical

Many chemicals are available for managing volunteer potatoes (Table 2). Use of herbicides and soil fumigants can assist in management of the volunteer potato problem and should be considered in conjunction with other management techniques as part of an integrated weed management plan. Some herbicides may restrict future cropping or re-cropping options; they should be managed properly and according to the label guidelines. Always refer to herbicide labels and the PNW Weed Control Handbook to determine legal product use and application rates. Fumigation may have potential as a management tool to help suppress volunteer plant growth in less competitive crops with limited herbicide options such as onions, carrots, and peas.

The cost of fumigation limits its use to high value crops and is normally used to help suppress nematodes and other soil pathogens. A fall or spring soil application of 1,3-D (Telone®) shanked in plus sprinkler-applied metham sodium (Vapam®) on commercial fields reduced viable tubers by approximately 75 percent (Boydston and Williams). Metham sodium is commonly used in the PNW through center pivot irrigation. It can reduce populations of annual weeds up to 95 percent. Results from soil fumigation can be variable due to nonuniformity of fumigant application, uneven distribution of fumigant in the soil profile, and use of sublethal doses.

During winters where soil temperatures were low enough to kill tubers in the top four inches of the soil, a shanked application of 1,3-D was more effective at managing deeper tubers than sprinkler-applied metham sodium (Boydston and Williams). Conversely, sprinkler-applied metham sodium may be more effective in years when a killing frost does not occur and shallower tubers could potentially be suppressed by a late summer or late spring application. Both dormant and nondormant tubers have been shown to be susceptible to metham sodium, but further research is needed on this topic. The effect of soil moisture and temperature on fumigant dose is not well defined but has been shown to have potential for refining fumigant application timings to obtain suppression. Corn has several herbicides labeled for management of volunteer potato growth and daughter tuber formation. Applications of atrazine followed by a post application of dicamba and 2,4-D effectively suppress volunteer potatoes (Boydston). The disadvantage of using atrazine is the limit it poses for the following crop. Another effective combination is a postemergence application of fluroxypyr (Starane®) following a pre-emergence application of atrazine.

These treatments can be further improved in combination with cultivation, dammer-diker operations, or a late season Colorado potato beetle defoliation.

Boydston's study showed a single mid-postemergence application of carfentrazoneethyl (Aim®) plus dicamba (Clarity®) reduced daughter tuber weights by 76 to 96 percent. Carfentrazone, which acts primarily by contact activity, requires multiple applications to manage regrowth of new potato shoots when used alone. This study also showed that the combination treatment had similar corn yield compared to the hand-weeded check. Mesotrione (Callisto®) is a new product that effectively controls or suppresses volunteer potatoes and is registered for use in field corn.

Volunteer potato control, tuber production, and corn yield were evaluated following treatment with several herbicides and herbicide combinations (Boydston and Williams). Mesotrione greatly reduced both the number of daughter tubers produced as well as tuber biomass compared to the other postemergence herbicide treatments tested. Mesotrione applied postemergence at the time of early potato tuber initiation reduced daughter tuber number and weight by 95 and 99 percent, respectively, compared to non-treated checks. The few daughter tubers that were formed often produced weakened sprouts, and sprouts were slow to emerge the following spring.

The corn yield of all the mesotrione treated plots was equal to the hand-weeded check. Some crop rotations may be restricted with the use of mesotrione. Sensitive crops include peas, beans, and alfalfa. Before using mesotrione, please refer to the manufacturer's label for application instructions as well as more information on rotation restrictions. In late-planted crops or in no-till corn, glyphosate applications can be used to manage early emerging volunteer plants. An application made pre-plant will kill the shoots of emerged potatoes, but new shoots frequently emerge two to three weeks later.

Consequently, subsequent methods will have to be used to adequately manage daughter tuber production. Maximizing shoot emergence prior to glyphosate application results in less potato regrowth following herbicide application than with earlier herbicide application. Onions are another rotation crop with limited options for managing volunteer potatoes. Cultivation alone has shown some level of effectiveness, but does not remove volunteer potatoes in the plant row. Multiple applications of oxyfluorfen (Goal®) in combination with cultivation has been shown to maintain onion yield (Boydston and Seymour). Applications of oxyfluorfen at the two-, three-, and four- to five-leaf stages of onions, followed by a cultivation after each application, reduced daughter tuber weight 69 to 96 percent and tuber number 32 to 86 percent compared with cultivation alone. This labeled application in onions proved to be quite beneficial. Other herbicides tested were effective in suppressing volunteer plants, but were phytotoxic to onions and reduced yield. Cereal crops are often grown in rotation following a potato crop.

For those volunteer plants present in a wheat crop there are limited chemical management options. Fall-planted cereals delay the emergence of volunteer potatoes in the spring. Fluroxypyr and metsulfuronmethyl can suppress emerged potato shoots, but applications must be made at specific stages of cereal growth, during which potato shoots have often not yet emerged.

A pre-harvest application of glyphosate can be applied to wheat (except seed wheat) at 30 percent or less grain moisture (hard dough stage) for management of volunteer potatoes. However, volunteer potatoes are usually water stressed at this time, below the wheat canopy, and have already formed daughter tubers, reducing the effectiveness of this option. Glyphosate can also be applied post harvest. Oregon studies have shown that an application of glyphosate to volunteer potato plants when the tubers are over two ounces can reduce tuber viability. Results from a glyphosate application at these times can vary, depending on the size of the volunteer plants and tubers, the condition of the plants, and the moisture availability. It may be necessary to make multiple applications to achieve adequate suppression.

7

Manure and Fertilizer

EFFECT OF COMPOST AND ANIMAL MANURE WITH PHOSPHORUS AND ZINC FERTILIZER ON YIELD OF SEED POTATOES

Environmental problems caused by irregular application of chemical fertilizers, inappropriate energy production methods and excessive consumption costs have all had harmful effects on biological cycles and destroyed farming stability systems; these factors altogether encourage the application of bio fertilizers. Sustainability and safety of food production associated with environmental protection and fair socio-economic interactions in the societies involved are the most important debates in many multidisciplinary issues of agriculture, ecology and environmental sciences.

These topics have attracted attention of researchers, farmers, policymakers and other stakeholders worldwide.Since soil as a living entity is the basis of management in sustainable food production, enhancing soil productivity through utilization of organic fertilizers has been gaining significant importance in organic food production. These organic sources have a profound effect on crop yield and quality. In organic farming composts, organic manures and their extracts are used for improving soil fertility and in combating pests and diseases. Organic manures and composts have been found to have a direct anti disease effect by stimulating competing micro organisms and also by inducing resistance to plant diseases.

However, there are other contradicting evidences indicating the reverse impact of using these sources. The use of fertilizers and manure to enhance soil fertility and hence crop yield improvement is a traditional method being used quite long a go by farmers. On the other hand the use of compost on agricultural land can be important in products based on the principles of sustainable agriculture.

Use of municipal compost affect on the economic and environmental factors, such as reducing transport costs and bury it, to support environmental laws, reduce the use of inorganic fertilizers and improved soils crop characteristics. The compost enriched with chemical fertilizers in the field the ability to access the elements of high consumption caused by the products and can be increased ability to production and soil fertility. Positive role of compost application has been reported in many crops, garden and pasture.The application of compost increased microbial activity, nitrogen concentration and grain yield. Use of compost in the Mediterranean semi-arid lands increased nitrogen, phosphorus, potassium and organic carbon content in the rhizosphere.

Application of manure in the soil causes soil hollow, increased water holding capacity in soil and improved it physical properties, besides increasing soil fertility, crop growth and thus the water use efficiency. The application of compost improved seed germination and yield dry matter as compared to compost-free treatments.

Phosphorus is the second most important macro nutrient after nitrogen that plays significant role in physiological and biochemical reactions such as photosynthesis and transfer characteristics. Phosphorus fertilizers and manure in the soil increased phosphorus uptake by plants, through favoring production of carbonic acid, the acid that increases solubility of phosphate compounds in calcareous soils.

Zinc is also one of the essential micronutrients required for crop growth since it is an important component carbonic anhydrase enzyme which is present in all photosynthetic tissues, and is also re quired for chlorophyll biosynthesis. The purpose of the present study was to determine the effects of compost, animal manure, phosphorus and Zinc fertilizer on seed potatoes production.

ROLE OF MANURE

Maintaining and improving soil quality in vegetable and potato systems is a difficult task. Heavy equipment can cause compaction, crop residues tend to be minimal in these rotations, and sandy soils generally have low organic matter levels. Degraded soils can be revitalized by adding manure or compost. To minimize costs while enhancing organic inputs, a combination of manure and cover crops is recommended to rebuild soils.

Both short-term yield benefits and long-term soil building can occur through the addition of organic matter. Recent studies in Michigan suggest that application of poultry manure with a reduced fertilizer rate can enhance potato tuber yields by 30 to 60 cwt/acre at some sites.

The mechanism is under study but may be related to any of the following: enhancement of soil microbial activity, the active organic matter fraction or enhanced nutrient supply. Other benefits take time to accrue and are generally associated with manures that contain large amounts of organic materials, such as dairy manure or manure from straw-bedded livestock. Soil quality benefits include increased water-holding capacity, improved soil structure and nutrient buffering ability. An increase in soil organic matter of 0.5 percent will increase water-and nutrient-holding capacity by about 10 percent.

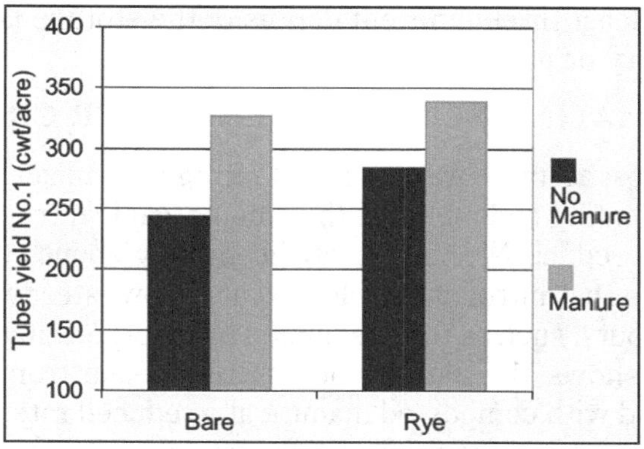

Fig. Chip potato response to 160 lb N/acre fertilizer plus 5,000 lb/acre aged poultry manure, supplying about 40 lb N/acre. The soil was left bare in the winter or a rye cover crop was used.

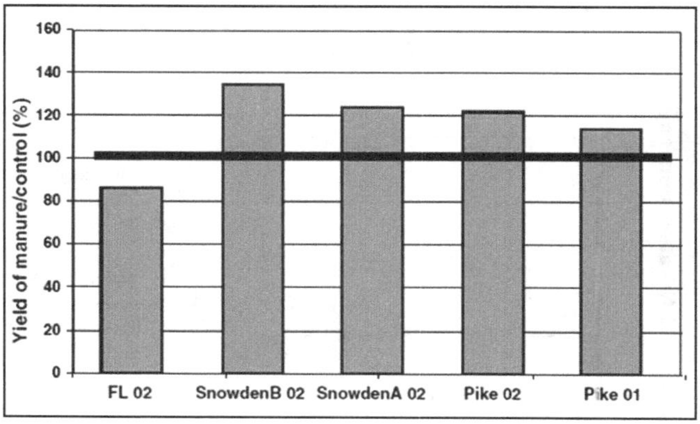

Fig. Summary of potato yield response as a percentage of control for experiments with five chip potato varieties conducted in 2001 and 2002

Benefits and potential negatives associated with using manure depend on the type of manure, how it is stored and the timing of application. It is highly advisable to obtain a laboratory analysis to judge the quality of the manure. Your county MSU Extension office has a list of manure testing laboratories.

MANURE ANALYSIS

Nitrogen and phosphorus content, pH and the amount of organic matter added are all useful measurements that provide insight into the quality of the manure. Total N and inorganic N are particularly useful for estimating N availability. Phosphorus builds in the soil, and the amount added with manure should be accounted for and subtracted from fertilizer applied. A sample is only as good as the sampling technique used, and it is important to obtain a representative sample from the manure storage facility. Your county MSU Extension office can provide guidance, or check with the testing laboratory to determine a recommended sampling procedure. One useful method is to obtain a sample from at least three different depths of the storage facility, preferably just after the slurry or pile has been well mixed.

ESTIMATING NUTRIENT FERTILIZER CREDIT

One of the most accurate ways to adjust nitrogen fertilizer rates is to sample soil just before planting and measure the soil inorganic nitrogen. If soil nitrate is above 20 ppm, reduce N fertilizer application by about 60 lb/acre. Recent research indicates that nitrogen is released at a slow, steady rate from mixed quality organic inputs, such as those supplied by incorporating cover crops plus manure. Figure shows the pattern of nitrate release from rye cover crop residues combined with composted manure at a reduced rate of fertilizer. This combined treatment supported the highest potato yields and the lowest fertilizer cost.

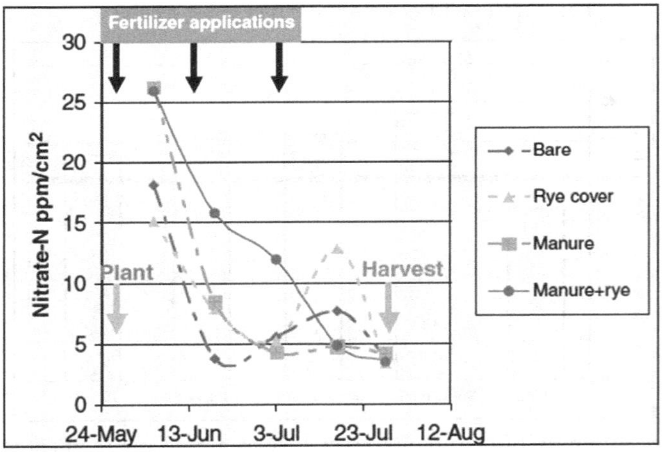

Fig. Soil nitrate measurements indicate availability of nitrogen (N) from 5,000 lb/acre of poultry compost applied to a bare or a winter rye cover crop-potato rotation, where N fertilizer was applied in three splits as indicated by the arrows. Total N availability was adjusted to 180 lb/acre for each treatment.

TIMING OF MANURE APPLICATION

There are trade-offs to be considered in timing manure applications. Manure application in the fall is easier to manage to obtain nutrient release in the spring when crop nutrient demand is high. Manure applied in the spring has the advantage of not being vulnerable to nutrient loss over the winter, but it may be a challenge to estimate fertilizer credits accurately. Applying manure to soil with wheat residues or a living cover crop helps optimize nutrient efficiency and benefits from manure. Another timing question is whether to apply manure ahead of or after fumigation. The lowest risk approach is to apply manure in the fall, just ahead of fumigation. Further research is required because it is also possible that benefits can be obtained from applying manure after fumigation. Adding manure directly before planting can increase risk of disease, such as common scab in potato. Consider carefully manure effects and disease potential before incorporating manure in the spring. Composted manure will generally be much safer and reduce disease risk.

Table. Positive and negative aspects of fall vs. spring manure application.

Fall application		Spring application	
Positive	*Negative*	*Positive*	*Negative*
Easier to estimate nutrient availability	Potential for N, P loss over winter	Less opportunity for nutrient loss	Nutrient release may be slow if temperatures are cool.
Lower disease risk	Bare soil after manure is applied	Availability from livestock producers	Disease risk may be higher

MANAGEMENT STRATEGIES

To optimize benefits from manure, growers need to consider both timing and placement. Improved management options include the following:

- Broadcast manure on a forage crop or a cover crop before production of vegetables or potatoes in a rotation sequence.
- Apply manure in the fall through injection or spreading followed by an incorporation operation, then plant a winter cover crop such as rye or wheat that can take up any nutrients released over the winter.
- Apply manure to a field with no-till wheat stubble or significant amounts of other crop residue remaining on the surface.
- Because manure is a heterogeneous substance and does not apply smoothly, double spreading (30-foot instead of 60-foot centers, for example) or applying the manure first in one direction and then in the other will improve uniformity of application.
- Calibrating the manure spreader is the foundation of good manure management and accurate estimation of nutrient supply.
- Apply poultry manure or slurries for short-term benefits, but do not expect long-term soil building unless manure with significant amounts of organic matter (e.g., straw-filled or solid dairy manure) is applied.

MANURE EFFECTS ON CROP DISEASE

In the past, manure has been associated with soilborne diseases such as common scab. Tuber soft rots also may be enhanced by applying manure, particularly large amounts, such as 20,000 lb/acre. Manure management and quality have changed, however. Straw bedding and solid material is no longer common. Lagoon storage of slurries is becoming the norm. Recent research suggests that acidic slurry manure does not consistently enhance the risk of common scab and may occasionally suppress it. To reduce disease risk, however, it is still highly advisable to use scab-resistant potato varieties and to apply manure in the fall rather than the spring.Recent research from the Great Lakes region indicates that two or more years of applying composted poultry manure, swine manure or composted paper mill residuals can suppress soilborne diseases in some cases. At some sites, root rot disease was suppressed and a healthy crop of snap beans, cucumbers or potatoes was produced. Further research is required to improve knowledge of how manure quality and quantity interact with site and crop type to influence disease management. The bottom line for healthy crops is that manure and compost should be applied with sufficient time for decomposition to occur. Ideally, these organic inputs should be applied to crops grown in rotation with vegetables, such as field crops, forages or cover crops.

FOOD SAFETY

It is important to reduce any risk of contamination from manure to prevent food-borne illness from E. coli or other bacteria. General guidelines are available in "Food Safety Begins on the Farm". They are summarized here, though the full text is recommended:

- Manure should never be used as a sidedress. It should be applied at least 120 days before a crop is harvested.
- Manure should be stored for 60 to 90 days before application, if possible.
- Well-composted manure is preferred, in particular when a 120-day waiting period is not feasible.

Fig. To build soil quality requires high levels of organic matter inputs.

USING GREEN MANURES IN POTATO CROPPING SYSTEMS

WHY USE GREEN MANURES?

Potato producers are using green manures to produce better crops by improving the quality of their soils. The physical, biological, and some chemical characteristics of soil may be improved by green manures. Physical characteristics such as water infiltration rate, tilth, water holding capacity, and aeration, are generally improved by the addition of organic matter to the soil, be it via manure, green manure, compost, or crop residues. This can lead to the growth of larger, healthier root systems which help plants better handle stress. The chemical properties of a soil can be improved by increasing nutrient and organic matter levels. This, too, comes from organic amendments to the soil. The biological characteristics of a soil, such as biomass, biological activity, and biodiversity, can also be improved through green manures. These changes in the soil's biology provide the shortterm economic incentive to use green manure crops in potato cropping systems, especially for soilborne pest management. Fungal and bacterial diseases, nematodes, and weeds can all be reduced by using a green manure crop. When used in certain cropping systems, green manure crops have been able to replace expensive fumigants. However, the degree and duration of these beneficial effects depend on many factors, such as soil texture, climate, tillage practices, and crop rotation. Therefore, the benefits of green manures may differ between systems.

HOW GREEN MANURES HELP MANAGE PESTS

The effects of green manures on soilborne pests are the results of several interacting mechanisms. These mechanisms take place in the complex environment of the soil where it is difficult to measure specific biological processes. It is not yet possible to say which mechanism is most important or how each works in conjunction with the other—we can only deduce which mechanisms may be at work. Still, it is beneficial to review these mechanisms and the strategies you can use to enhance their effects in your system.

Crop Rotation

Before advances in soil microbiology, many green manure and cover crop effects were combined under crop rotation. Crop rotation reduces pest problems by changing the environmental conditions in the field. Each pest has a set of conditions it prefers. If pests are allowed to have their favored set of conditions for too long, they multiply rapidly and give us problems. In general, rotating crops with different planting dates, different growing habits (annual vs. perennial, tall vs. short, fibrous vs. tap roots) or different susceptibility to pests (grasses vs. broadleaves) prevents any one pest from becoming a problem.

Strategy

Rotate crops that are as different from one another as possible, and usually, the longer the rotation, the better the pest control. With green manures, grow a crop that is not a host to the pests that affect your main crops.

Competitive Exclusion

A second mechanism of some green manure crops has been termed competitive exclusion. This is the mechanism put forward by scientists to account for suppression of *Verticillium dahliae*, the cause of verticillium wilt in potatoes, after a green manure crop. They have observed that when certain green manures (barley, mustard, rapeseed, sudangrass, and sweet corn) are incorporated before a potato crop, the level of infection by *Verticillium* is low even with high levels of the fungus still in the field. The green manure serves as an energy source for beneficial microorganisms. It is suspected that these beneficials out-compete *Verticillium* for energy and increase in number. Then, after the potato crop is planted, they exclude *Verticillium* from the area along the potato roots. This area, called the rhizosphere, is the only place where *Verticillium* can infect potato plants. Thus, exclusion prevents infection. There are probably other mechanisms at work resulting from the incorporation of a green manure. These may include predation, parasitism, and the interference of chemical signals between pathogens and plants. However, little is known about the specifics of these mechanisms and how they might affect control of soilborne pests.

Strategy

Produce a good green manure crop. The more biomass produced, the better, as long as it does not affect your ability to establish the following cash crop. This is where fall incorporation can be helpful; large amounts of biomass have longer time to break down before you need to plant your potato crop. However, in some conditions, fall incorporation may also increase wind erosion and winter leaching of nitrate. For this purpose, dried crop residues do not work as well as fresh plants. Therefore, incorporate the green manure crop while it is still green.

Biofumigation

This term was coined to describe the effects of Brassica rotation crops or green manures on soilborne pests. Brassica crops such as rapeseed and mustard contain biologically active chemicals, called glucosinolates. In the soil, certain glucosinolates in the roots of rotation crops, or in the roots, stems, and leaves of green manures, break down into isothiocyanates and other chemicals. ITCs are known to kill or suppress some soilborne diseases, nematodes, and weed seeds. There are many types of glucosinolates, some of which produce different types of ITCs, which vary in their toxicity to different pests. Methyl ITC is the active chemical produced when metham sodium, a common synthetic fumigant, is applied to the soil; hence the name biofumigation when ITCs are produced by plants. In the laboratory, these chemicals have suppressed growth of silver scurf, white rot, powdery scab, and pink rot. Field tests have shown mixed results and research is ongoing. Sorghum-Sudangrass and Sudangrass green manures have been shown to be effective against root-knot nematodes by a very similar mechanism. These crops produce dhurrin, which like glucosinolates, produce a toxic compound, in this case hydrogen cyanide (HCN), when incorporated into the soil.

Strategy

First, select species and varieties that produce large amounts of biomass with a high concentration of glucosinolates. Generally, the concentration of glucosinolates peaks just before flowering. However, biomass continues to increase until the plants begin to dry. The time of incorporation for maximum biofumigation is not yet known. Second, to produce ITCs, the glucosinolates must be exposed to specific enzymes, which are normally separated from the glucosinolates in the plants. This is also true for producing HCN from dhurrin in sudangrass. Current practice is to chop the green manure before incorporating to insure that this mixing occurs. A high-speed flail chopper, such as those used by some grass seed and asparagus growers, may be the best implement for this. Finally, ITC production is greater in wet soils than in dry soils, so if possible, irrigate following incorporation.

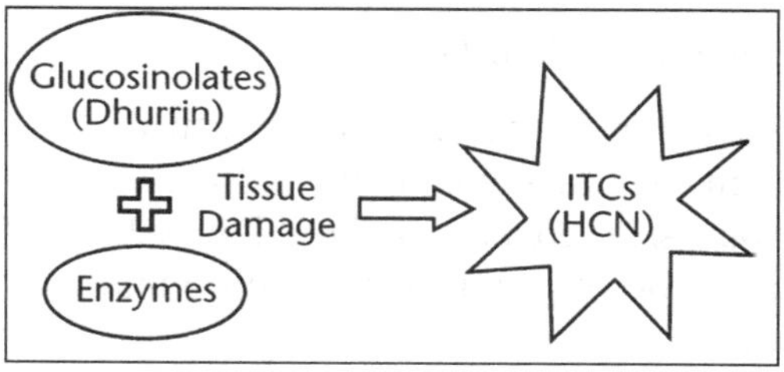

Fig. Sudangrass chemicals shown in parentheses.

Induced Resistance

This mechanism is also called organic mattermediated disease suppression. Similar to competitive exclusion, it occurs when beneficial organisms are stimulated by the addition of organic matter to the soil. Some of these beneficials then secrete chemicals that come in contact with plant roots and induce resistance in the plant. In some cases, resistance may be induced to insect-borne diseases. This is an effect similar to that of the harpin protein, which is sold in a product called Messenger®. Because we know so little about this mechanism's components, there is no clear strategy for enhancement. Incorporating a green manure crop may or may not provide the active organic matter in the soil that this mechanism depends upon. Research is ongoing.

INTERACTION OF PEST CONTROL MECHANISMS

We have identified four potential mechanisms that could reduce soilborne pest problems. However, you may have noticed that these mechanisms act in ways that may counteract each other. For instance, biofumigation is killing microorganisms while competitive exclusion depends upon increasing the number of certain microorganisms. What really happens in the soil is complex and presently beyond our ability to observe. That is the bad news. The good news is that the quality and quantity of your crop yield, balanced by the cost of production, will give you a good idea of the benefits of green manures in your cropping system.

FACTORS TO CONSIDER

Goals for a Green manure

Before you add a green manure crop to your rotation, you should decide what you want to achieve. It may be that you want to control a certain nematode, a disease, a problem weed, or just improve the soil's tilth. Once you have decided on a primary goal, then your management decisions should be made to maximize the effects that take you towards that goal. Often you may have secondary goals, but because there are different ways to manage green manure crops, you should always have your primary goal in mind.

Your Field's Combination of Ppests

Goal setting should consider the unique mixture of crops, soils, and pests on your farm or even in each field you manage. Because certain green manures will be better in certain situations, you should investigate the advantages and disadvantages of each type of green manure crop and try those that seem to best fit your situation.

Green manure Management

Success or failure in reaping benefits from a green manure crop can be a matter of management details. Factors such as planting method, planting date, seeding rate, fertilization, weed control, and incorporation method and timing can all help or hinder your goals. Before you buy any seed, make sure you have all the information available on the green manure you have chosen.

Cost

Green manure crops, in certain situations, can be costly. However, green manures produced by using the cheapest seed, residual soil nutrients, and minimal irrigation will not normally produce satisfactory results, especially in terms of pest management.

When looking at the cost of green manures and comparing them with the benefits, it is important to consider all the benefits. While it is relatively easy to figure the benefits if you can reduce or eliminate pesticide applications, it is far more difficult to estimate the benefits of improving soil physical characteristics, or other long-term benefits. Also, remember that some of the money you spend on a green manure crop would also have been spent if you did not grow the green manure. This is the case with fertilizer applied to green manures as much of it will be recycled into following crops, thus reducing the amount of fertilizer needed.

Management and Labor Requirements

When considering a green manure, take into account the time and labor required to obtain good results. Will you have these resources available when they are needed? Do you have other activities at that time of year that may keep you from paying due attention to the green manure crop? You may have to supply additional machinery or labor if you want to successfully produce a green manure crop.

Short-term vs. Long-term Benefits

While many of the pest management benefits may come soon after producing a green manure crop, other benefits will not be evident for years. If you regularly use green manure crops, you may find that the benefits increase every year. If you only grow a green manure crop every four or five years, you may find that the short-term benefits are all you see. In the short term, adding a cover crop to a cropping system will impact nitrogen dynamics. The N released from the crop residues will affect the timing and amount of fertilization needed. Additional soil and/or tissue sampling may be needed to optimize this nitrogen.

Other Cropping System Factors

You can affect the success of your green manure cropping through other management decisions. In general, soil quality will improve or be maintained if you:

- Minimize tillage
- Avoid over watering
- Avoid soil compaction
- Prevent wind and water erosion

Changing System Components

Remember that you are working within a system and changing the components of a system may change the outcomes. For instance, one successful cropping system has a mustard green manure crop following wheat. In this system the mustard is planted without disturbing the wheat straw. In late October, both the mustard and wheat straw are incorporated together. You may want to change this system by incorporating the wheat straw before planting the mustard. This may seem like an insignificant change. However, it may be that the incorporation of the wheat straw with the mustard is important in the success of this system. By incorporating the wheat straw you might increase the weed pressure, increase the nitrogen requirements of growing the mustard, or increase the risk of nitrate leaching and wind erosion in the following winter and spring. This should not prevent you from developing your own system, but be aware of the possible complications when changing system components.

GETTING STARTED WITH GREEN MANURES

On-farm testing is a good way to evaluate green manures when done correctly. However, a single side-by-side comparison, although easy to conduct, will not tell you if your observations were the result of the practices you were comparing or the result of other varying conditions. For best results:

- Start with a small part of a larger field
- Leave areas that are managed as normal
- Use replication and randomization

Call your local extension office for help and resources to conduct on-farm tests.

KRAAL MANURE AS FERTILIZER

PLANTS NEED WATER AND NUTRIENTS TO GROW WELL

- All plants need water and nutrients to grow well.
- Water and nutrients are extracted from the soil by the roots of plants.
- Soil water removed by plants is regularly replaced by rain or by means of irrigation.
- In forests and grasslands nutrients taken up by plants are returned to the soil mainly through a process of decaying litter.
- Dead plant material such as leaves and stems accumulate on the soil surface, where it decays or decomposes. The nutrients contained in this plant litter is returned to the soil.
- On cropped land most plant material and the nutrients contained in it is removed from the land in the form of grain, leaves or other plant parts.
- In order to maintain healthy growth of crops it is necessary to replace the nutrients that have been removed by the crops. Replacement of nutrients is achieved by means of fertilising the soil.

CHEMICAL FERTILISERS VERSUS MANURE

In South Africa most large-scale farmers use chemical fertilisers on their soils. These fertilisers are bought in bags from fertiliser suppliers. In the rural areas chemical fertilisers are not easy to come by. One often has to travel to town to find a supplier. Chemical fertilisers are bulky and heavy. Getting these from town to your village or farm is cumbersome and you may have to pay for the transport. Chemical fertilisers are also too expensive for many households. Yet, without fertilising your garden or field soil, the crops you plant will not grow very well and you will most probably be disappointed with the yields, especially considering the amount of work you have invested in tending your crops. There are, however, alternatives to the use of chemical fertilisers. One alternative is kraal manure. Before the introduction of chemical fertilisers, farmers all over the world, including those in South Africa, made use of manure to restore the fertility of their lands. Many farmers turned to chemical fertilisers because these fertilisers are more concentrated than manure and easier to handle with mechanised planting equipment. Another reason is that the quantity of manure available on large farms is usually insufficient to fertilise all the land that is being planted to crops. However, when cropping a small area only, the available supply of manure may be quite adequate and the work involved with transporting manure to the lands not too difficult to handle.

THE KRAAL AS A SOURCE OF MANURE

- In rural areas the most common source of manure is the kraal.
- The kraal is an enclosure in which livestock are kept during the night, providing protection against theft and wild animals.
- The excrement and urine of livestock in the kraal accumulate on the floor of the kraal forming a layer of manure.
- Manure is essentially organic material consisting of the residues of plants that were digested by the animals housed in the kraal.

WHICH NUTRIENTS ARE THE MOST IMPORTANT?

The main nutrients a plant needs are nitrogen, phosphorus and potassium.

Phosphorus

In most South African soils phosphorus is the most deficient nutrient. This means that if you would have to choose between the application of either nitrogen, phosphorus or potassium, your crop yields would show the largest increase when phosphorus is applied to the soil.

Nitrogen

Nitrogen is usually the second most deficient nutrient and once you have taken care of the phosphorus needs of your crops, you can expect an additional increase in yield by applying nitrogen.

Potassium

In many parts of the country the amount of potassium stored in the soil is considerable and applying potassium may not always result in higher yields. Over time, however, continuous cropping will also deplete the potassium reserve in the soil, making it necessary to fertilise the soil with this nutrient in order to maintain high yields. In high-rainfall areas or where the soils are sandy, the quantity of potassium stored in the soil is usually low and fertilising with potassium may be necessary from the start.

DOES MANURE CONTAIN ALL THE IMPORTANT NUTRIENTS?

In this booklet the focus is on the use of manure only. The strategy of combining manure with mineral fertiliser will therefore not be discussed.

- Manure contains all the nutrients a plant needs, but not always in the desired proportions.
- Chemical analysis of kraal manure shows it to be quite high in potassium and relatively low in phosphorus and nitrogen. That is unfortunate, because phosphorus and nitrogen are the two nutrients soils are most deficient in.
- There are two ways of addressing this problem.
  - One can apply enough manure to meet the nitrogen and phosphorus needs of the crops. This will result in the supply of more potassium to the soil than is needed, but it is not really harmful.
  - The alternative is to combine the use of manure with mineral fertilisers. By doing so the rates at which manure needs to be applied (amount of manure per unit area of land), is much reduced. However, most mineral fertilisers have to be purchased, requiring you to spend money.
- As was indicated earlier, the supply of phosphorus is most crucial. Some of the most common sources of mineral phosphorus are chemical phosphates, such as superphosphates, bonemeal and rock phosphate.
- Bonemeal and rock phosphates release phosphorus very slowly. The beneficial effects are not as immediate as those of chemical phosphates, but are longer lasting, therefore reducing the need for regular addition.
- Manure is an organic fertiliser and unlike in chemical fertilisers, the nutrients that are present in manure are not immediately available to plants.
- In order to release these nutrients and make them available to plants, the manure needs to be decomposed or broken down.
- In the soil manure is decomposed by very small organisms. These organisms are usually too small to be seen with the naked eye, but they are always present in the soil. The application of manure and other organic matter increases their numbers.
- The soil organisms break down manure rapidly when the soil is warm and moist.
- When the soil is dry, cold, or both, decomposition takes place at a much slower rate. Therefore, if you apply manure to your soil in winter, when the soil is cold and dry, little may happen to the manure. But as soon as the weather warms up and rain adds water to the soil,

the manure will be broken down rapidly and within a few days nutrients are released to the soil, where they can be reached and absorbed by plant roots.

- The release of nutrients occurs over quite a long time and during this period increasing quantities of plant nutrients become available to your crop.

ARE SOIL TESTS NECESSARY?

Many commercial farmers send their soils for testing to a laboratory. The technicians then determine the nutrient content of soils. When sending samples to a soil test laboratory farmers are asked which crops they intend to grow on the lands from which the samples were taken. This information and the results of the soil tests are used by expert advisors to formulate a fertilisation programme. Farmers will receive a letter informing them which fertilisers to apply and the rates at which the fertilisers need to be applied to ensure optimum growth of the crops and high yields. Having your soil tested is a good idea, but it does not come cheaply. A complete analysis of one soil sample may cost you more than R100 and many of you may not have that kind of money to spare. We will therefore assume that you will not send your soil to a laboratory for testing.

WHAT IS THE NUTRIENT CONTENT OF MANURE?

At the University of Fort Hare manure samples from many different kraals in central Eastern Cape were analysed. On average a fully heaped modestly compacted wheelbarrow load of kraal manure was found to contain 0,6 kg nitrogen, 0,2 kg phosphorus and 0,8 kg of potassium. The manure contained in the wheelbarrow had a mass of about 75 kg, comprising about 15 kg of water and 55 kg of dry manure. The wheelbarrow that was used, was the one commonly used by builders to mix cement. It is the type with the high sides, not the gardener's wheelbarrow with the low sides.

UNDERSTANDING THE FORMULAS USED IN DESCRIBING CHEMICAL FERTILISER MIXTURES

In South Africa chemical fertiliser mixtures are described by means of a formula. This formula is printed in bold on the bag containing the fertiliser mixture. An example of a formula of a fertiliser mixture is 3:2:1 component of the formula indicates the concentration of the combination of pure nitrogen, phosphorus and potassium present in the fertiliser expressed as a percentage. Practically, it means that in every 100 kg of 3:2:1 (25) fertiliser mixture there is 25 kg of pure nutrients consisting of a combination of nitrogen, phosphorus and potassium.

The higher the number in brackets, the more concentrated the fertiliser mixture will be. A 50 kg bag of 3:2:1 (30) fertiliser will contain more nutrients than a 50 kg bag of 3:2:1 (25) fertiliser and will, as a result, also cost more. The 3:2:1 component of the formula indicates the proportion of nitrogen, phosphorus and potassium in the fertiliser. In the case of fertiliser mixture 3:2:1 (25) the mixture contains 25 % pure nutrients in the proportion three parts nitrogen, two parts phosphorus and one part potassium.

A USEFUL CHEMICAL FORMULA FOR KRAAL MANURE

Kraal manure can be described by the same formula used to describe chemical fertiliser mixtures. In the central areas of the Eastern Cape the average formula for kraal manure was found to be 3:1:4 indicates that in every 100 kg of manure there is a total of 3 kg of pure nutrients, consisting of a combination of nitrogen, phosphorus and potassium. The 3:1:4 component of the formula indicates that in manure these nutrients occur in the ratio three parts nitrogen, one part phosphorus and four parts potassium.

From the (3) in the formula of kraal manure it is obvious that the concentration of nutrients in kraal manure is not very high. For example, when compared to the chemical fertiliser mixture 3:2:1 (30) the nutrients in kraal manure are ten times less concentrated. This means that you would have to apply 1 000 kg of kraal manure to supply the same amount of nutrients found in 100 kg of the chemical fertiliser mixture 3:2:1 (30).

IS THE NUTRIENT CONTENT OF KRAAL MANURE ALWAYS THE SAME?

The concentration of nutrients in kraal manure depends on a number of factors, but the most important one is soil content of the manure. The more soil the manure contains, the lower its nutrient concentration and the more you have to apply to supply the same amount of nutrients. In the central areas of the Eastern Cape, kraal manure was found to contain between 20 and 80 % soil by mass, causing the formula to range between 3:1:4 (7) where the soil content was low and 3:1:2 (2) where more soil was present.

For practical purposes all the recommendations in this booklet are based on kraal manure containing about 60 % of soil by mass for which the formula 3:1:4 (3) applies. When your manure contains little soil you can reduce the amount of manure recommended here, but there would be absolutely no harm in following the application rates recommended.

After all, to increase the nutrient content of your soil to levels harmful to crops would require very large applications of manure, of which there is no danger when following the recommendations in this booklet.

ESTIMATING THE SIZE OF YOUR FIELD OR GARDEN OR PLOT

- In order to be able to supply manure at the recommended rates (amount of manure per unit area of land), you need to know the area of your plot or field.

- You can estimate the size of your plot or field fairly accurately by pacing the length and width of your field. One pace is about the same length as a metre.
- Determine the length of your field by counting the number of paces it takes you to walk from one side of the field to the other along the length of the field (the longest side of the field).
- Do not forget to record the number of paces, so take along a piece of paper and a pen or pencil.
- The same procedure is followed to determine the width of the field (the shortest side of the field). You count the number of paces while walking from one corner of the field to the opposite corner along the short side of the field and record the number of paces it took you to get there.
- To determine the area of the field, you multiply the number of paces obtained for the length of the field with the number of paces counted for the width of the field. This will give you the approximate area of your field in square metre (m2).
- The procedure to be followed when determining the area of a square or rectangular plot.
- If the shape of the land is not rectangular or square you can still determine its area. In Figure the procedure to be followed for plots with constant width but differing length is shown. The same procedure applies for plots with constant length but differing width.
- Determine the width (W) of your plot by pacing across the plot at right angle to the field edges.
- Determine the length (L) of your plot by pacing along the length of the plot on both sides.

FERTILISING YOUR CROP WITH KRAAL MANURE

Once your soil is cured, you should keep it in a healthy and productive state. This is best achieved by supplying the field soil with the nutrients that will be removed by your next crop.

Do this before you plant the crop. Crops differ in the amount of nutrients they remove. Some crops remove nutrients in large amounts while others do not. Furthermore, the higher the yield of the crop, the more nutrients the crop removes from the soil.

USING A CART OR SLEDGE, HOW DO I DETERMINE THE APPLICATION RATE?

- In case you transport kraal manure with a cart or sledge, you will need to determine how many wheelbarrow loads it takes to fill the cart or sledge.

- To do this you empty wheelbarrows of kraal manure into the cart or sledge until it is full while counting the number of wheelbarrows it takes to fill it.

HOW AND WHEN SHOULD MANURE BE APPLIED?

- Kraal manure is usually applied during winter, which is the fallow period for most crops.
- The recommended quantity of manure is spread evenly over the plot and worked into the soil to a depth of 5 to 10 cm soon after application.
- Working manure into the soil can be done with a spade in small gardens or by plough in large fields.
- Applying manure during winter provides enough time for it to decompose before the crop is planted during spring or summer.
- In areas where crops are grown in winter, manure can be applied during late spring, at least one month before planting the crop.

LIME AND ASH IMPROVES THE EFFECT OF KRAAL MANURE ON ACID SOILS

When farming in an area where the rainfall is relatively high, it is most likely that your soil is acid. In acid soils nutrients are rendered unavailable, meaning that they are present in the soil, but plants are not able to extract them from the soil. Therefore, when you apply manure to an acid soil, the yield of your crop may not increase much, because the nutrients released from the manure are not available to your crop and remain fixed in the soil.

This problem can be addressed by applying lime to the soil. Large quantities of lime are required to reduce soil acidity effectively and you may have to apply 2 to 3 tons of lime per ha to neutralise soil acidity. This does not come cheaply, especially when you live in a remote area.

In case you cannot get hold of lime, or are unable to afford lime, you could make use of wood ash to improve the quality of your manure. Save the ash from cooking fires and mix it with the kraal manure before applying the manure to the plot or field. The wood ash will help in keeping the nutrients which are released from the manure available to your crop, at least for a short time. Mixing wood ash with your manure will improve the effectiveness of manure applications on acid soils and will give higher yields.

WEEDS ALSO GROW BETTER WHEN FERTILISED: CONTROL THEM REGULARLY!

- Crop yield does not only depend on the quantity of nutrients in the soil.
- In order to produce high yields crops need an adequate supply of water and should be kept free of weeds.

- Weeds are also plants and like crops they remove water and nutrients from the soil.
- When you allow weeds to grow among your crops, they will use most of the water and nutrients intended for your crops.
- Remember, weeds will also grow better when the soil has been made fertile by manure applications.
- Allowing weeds to grow among your crop will largely remove the benefit you expect to derive from fertilising your soil with kraal manure.
- The results obtained from growing a crop on fertilised soil where weeds are not eradicated will be very disappointing, because instead of harvesting food you will produce a beautiful crop of weeds and very little to feed your family or to sell on the market.

DISEASES AND PESTS

Even when you have supplied the required nutrients to your crop, made sure the plants are never short of water and kept the weeds in your plot under control, it is still possible that your crop could be destroyed by pests or diseases. The control of pests and diseases in crop production is a complex matter and a number of pest control strategies can be followed. We intend to publish information on appropriate plant pest control strategies for small-scale farmers, paying special attention to some of the most commonly grown crops.

Bibliography

A K Kolay: *Manures and Fertilizers*, Atlantic Publication, Delhi, 2007.

A. Daniel Hall: *Fertilizers and Manures*, Biotech Books, Delhi, 2002.

A.P. Jaiswal and Deo Kant Prasad: *Handbook of Soil Fertilizer and Manure*, Enkay Publishing House, Delhi, 2012.

Amrita Priya: *100 Ways to Cook Potatoes*, Jaico Publication, Jaipur, 2001.

Anshuman Mohan: *Potato Chips*, HarperCollins, 2010.

Arthur W. Gilbert, Mortier F. Barrus and Daniel Dean: *Growing and Breeding of Potatoes*, Asiatic Publication, Delhi, 2006.

B.C. Mazumdar: *Orchard Irrigation and Soil Management Practices*, Daya Publication, Delhi, 2004.

Benu Singh: *Soil Management*, Vista International Publication, Delhi, 2008.

Dev Raj: *Technological Innovations In Potato Processing*, Narendra Publication, Delhi, 2012.

E.J. Russell: *Handbook on Soils and Manures*, Discovery Publication, Delhi, 1999.

H.W. Dalzell, A.J. Biddlestone, K.R. Gray and K. Thurairajan: *Soil Management : Compost Production and Use in Tropical and Subtropical Environments*, Daya Publication, Delhi, 2007.

L K Tripathi: *Water and Soil Management*, ABD Publication, Delhi, 2005.

M S Shaktawat; Mohan Singh; N C Aery; B N Swami and M K Katewa: *Phosphate Rich Organic Manure : An Alternate to Phosphatic Fertilizers*, Himanshu Publication, Delhi, 2004.

N C Aery; N S Rathore; M K Katewa and M R Masih: *PROM (Phosphate Rich Organic Manure), Vol. I*, Himanshu Publication, Delhi, 2006.

Narendra Singh and Zakwan Ahmed: *Potato : Production, Storage and Marketing*, Satish Serial Publication, Delhi, 2010.

P K Behera: *Plant Compost-Manure and Agro-Chemicals Analysis : A Laboratory Manual (2 Vols-Set)*, Dominant Publication, Delhi, 2007.

Philippe Giordanenge, Charles Vincent and Andrei Alyokhin: *Insect Pests of Potato*, Bio-Green Elsevier, Delhi, 2015.

R C Mandal: *Tropical Root and Tuber Crops : Cassava (Tapioca), Sweet Potato, Aroids, Yams, Yam bean, Coleus*, Agrobios Publication, Delhi, 2006.

Rakesh Mani Sharma: *Compendium of Indian Potato Research 1992-2011 : A Bibliography: Vol. 3*, Surinder Paul and B.P. Singh, Biotech Books, Delhi, 2013.

S C Panda: *Soil Management and Organic Farming*, Agrobios Publication, Delhi, 2006.

S K Mathur B B S Kapoor: *Emerging Trends In Soil Management*, Madhu Publications, Delhi, 2007.

S.K. Singh and D.K. Singh: *Pest and Soil Management of Horticultural Crops*, Agrotech Publication, Delhi, 2007.

Sharad Chandra Yadav: *Potato : Production Processing and Marketing*, Biotech Books, Delhi, 2008.

Uma Sah, K.D. Kokate, B.P. Singh, V.K. Gupta and S.K. Dubey: *Sustainable Potato Production in Hilly Region of India*, Biotech Books, Delhi, 2014.

Index